W0081871

Algorithmic Governance

Ignas Kalpokas

Algorithmic Governance

Politics and Law in the Post-Human Era

Ignas Kalpokas
Department of Public Communication
Vytautas Magnus University
Kaunas, Lithuania

ISBN 978-3-030-31921-2   ISBN 978-3-030-31922-9   (eBook)
https://doi.org/10.1007/978-3-030-31922-9

© The Editor(s) (if applicable) and The Author(s), under exclusive license to Springer
Nature Switzerland AG 2019
This work is subject to copyright. All rights are solely and exclusively licensed by the
Publisher, whether the whole or part of the material is concerned, specifically the rights
of translation, reprinting, reuse of illustrations, recitation, broadcasting, reproduction
on microfilms or in any other physical way, and transmission or information storage and
retrieval, electronic adaptation, computer software, or by similar or dissimilar methodology
now known or hereafter developed.
The use of general descriptive names, registered names, trademarks, service marks, etc. in this
publication does not imply, even in the absence of a specific statement, that such names are
exempt from the relevant protective laws and regulations and therefore free for general use.
The publisher, the authors and the editors are safe to assume that the advice and
information in this book are believed to be true and accurate at the date of publication.
Neither the publisher nor the authors or the editors give a warranty, expressed or implied,
with respect to the material contained herein or for any errors or omissions that may have
been made. The publisher remains neutral with regard to jurisdictional claims in published
maps and institutional affiliations.

Cover illustration: © John Rawsterne/patternhead.com

This Palgrave Pivot imprint is published by the registered company Springer Nature
Switzerland AG
The registered company address is: Gewerbestrasse 11, 6330 Cham, Switzerland

*This book is dedicated to the memory of Ona Petroševičiūtė (1932–2019)
who was my ultimate inspiration.*

ACKNOWLEDGEMENTS

This book would not have been written without my wife Julija. Not only due to her unwavering support despite my awkward teaching and writing schedule but also due to her substantial impact on the ideas that went into this book. It was she who sparked my interest in legal and regulatory matters. It was she who inspired my interest in regulation through code by introducing me to the writings of Lawrence Lessig. It was she who first encouraged me to think about algorithmic governance from a legitimacy and human rights perspective—these considerations ended up underpinning Chapter 6 in this book (any inaccuracies are, of course, my own). And my first foray into publishing on regulation in the digital sphere was an article we co-authored back in 2012. I am grateful to her from the bottom of my heart.

In addition, I would like to thank my students at LCC International University, particularly Yuliia Diadiuk, Amy Duckworth, Jesse Diblasi, Valeriia Petrechkiv, and Danielė Tumosaitė, for their comments on the draft manuscript and advice on making it more accessible. The interest that they and other students took in my manuscript was inspiring beyond measure.

Finally, for a book focused on agency being shared between human and non-human actors, it must be admitted that a significant role in the writing process was performed by algorithms, primarily those that run the searches on Google Books and Google Scholar as well as on publisher-specific databases as well as Amazon's book recommendation algorithm (and the content recommendation algorithm on YouTube, which was sometimes necessary to keep the children entertained). Hence, they all must be seen as indispensable agents in the writing of this book. Also acknowledged should be the agency of caffeine, in the form of both coffee and energy drinks, which fueled my writing. Inter-city coaches and children's playgrounds, meanwhile, served as important backdrops for research. And while the list could go on and on, I hope the preceding does help illustrate that there is more to agency than just the author. If any of those factors (or those I omitted) did not exist or was different, a book you are reading may have been a different version of itself.

As a separate note of appreciation, I would also like to thank Ambra Finotello for her enthusiasm about the project and Anne-Kathrin Birchley-Brun for her help and support during the submission process. I am grateful for the effort that they and everybody else at Palgrave Macmillan have put into making this book happen.

CONTENTS

CONTENTS

CHAPTER 1

Introduction: I'll Be Watching You...

Abstract Technological changes, particularly those related to the collection, analysis, and application of data, have significantly disrupted the personal, public, and economic domains. With both human persons and their environments being knowable to an unprecedented extent, and trends, correlations and future trends derivable from the accumulated data, a new mode of economic organisation, usually referred to as the platform economy, has taken hold. However, the internal logic of this economy is impossible to contain in its own domain and has, therefore, also spread into the governance of everyday life, not least because an ever-growing part of everyday life is lived *through* such platforms. Hence, this introductory chapter sets out the context for the book by delving into today's digital-first transformations.

Keywords Platform economy · Sensors · Data · Governance · Commodification

In my childhood, a radio station used to be on all the time, constantly broadcasting hits from the 60s, 70s, and 80s. In the post-Soviet years, that was probably a matter of a previously scarce resource (Western songs from my parents' teenage years and youth) suddenly becoming widespread. Among them, there was one that I had always found strange. Even today it sounds to me like the ultimate stalkers' anthem. The song is, of course, *Every Breath You Take* by The Police. With lyrics like '*Every*

© The Author(s) 2019

I. Kalpokas, *Algorithmic Governance*,

https://doi.org/10.1007/978-3-030-31922-9_1

breath you take / Every move you make […] Every step you take / I'll be watching you', it can certainly sound sinister. However, this mantra is also evocative to anyone concerned about privacy and datafication in today's digital world. Indeed, by now, as Cowley (2019: 96) attests, 'it seems uncontroversial – even banal – to suggest that we have become reflexively aware that our actions, from the moment we wake up, are digitally mediated'. However, they are much more than just mediated—they are permanently measured and quantified, thus giving rise to comparison and competition, prognostication, and surreptitious adaptation of the digital architecture of the everyday. Hence, the online platforms behind these processes must be seen as not only reflecting but, even more so, *producing* 'the social structures we live in' (van Dijck et al. 2018: 2), thereby disrupting traditional patterns of life and governance.

Algorithmic governance—the increasingly prevalent form of governance in this digital world—is characterised by its tackling of problems through 'their effects rather than their causation': Instead of disentangling the multiplicity of causal relationships and getting to the root of every matter, this form of governance is intent on collecting as much data as possible in order to establish robust correlations; in other words, instead of decoding underlying essences, this mode of governance works by way of establishing connections, patterns, and, no less crucially, predictions (Chandler 2019: 24–29). These can be subsequently worked on and turned into algorithmically devised courses of action, changes in the digital architecture of our everyday environment, or nudging strategies. Notably, though, this attitude that prides itself on replacing causes with trends also has the effect of altering the place of human persons, effectively objectifying and commodifying them, turning them into data generators where the data footprint is all that matters and is *taken for* the person.

The shift towards algorithmic governance has happened courtesy to five major recent developments: 'data, algorithms, networks, the cloud, and exponentially improving hardware' (McAffee and Brynjolfsson 2017: 95). The abundance of data from sensors embedded almost literally everywhere has allowed access to the world and the individuals inhabiting it in unprecedented detail; algorithms have enabled the analysis of and pattern recognition in this deluge of otherwise raw material; networks have enabled cheap, instant, and virtually ubiquitous transmission of both data and the results of their analysis; the cloud has enabled vast and flexible storage and computing space to both perform data-related tasks and develop new, more potent, algorithms; finally improvements in hardware have added sheer power to the capture and analysis of data. All of these developments have severely disrupted (and, indeed, transformed) not only private life and the economy but also governance processes.

However, it is not only the matter of data capture, analysis, and use that is at stake. In fact, a crucial theme and concern of this book is how new data-based techniques of governance are changing the very matter of being human. Agency suddenly becomes debatable: after all, as Mau (2019: 3–4) stresses, '[i]f everything we do and every step we take in life are tracked, registered and fed into evaluation systems, then we lose the freedom to act independently of the behavioural and performance expectations embodied in those systems'. It also must be kept in mind that whereas various forms of regulation and enforcement of expectations traditionally used to have a *public* nature—i.e. were typically adopted, promulgated, and enforced by a public authority of some sort, the algorithmic governance regimes of today have a distinctly *private* character, being part of what is increasingly referred to as the platform economy or platform capitalism (see e.g. Srnicek 2017) or 'digital capitalism/Big Data capitalism' (Fuchs and Chandler 2019). In a broad sense, it can be defined as 'a system in which a small group of powerful technology firms have vertically integrated a vast range of services and functions that they then provide to others' (Hill 2019: 3). But perhaps more than integration, this system is characterised by a specific and distinct logic: provision of services or connection of service providers and users in exchange for data and the use of such data to better target and more efficiently discipline both users and providers. In fact, the efficiency aspect is paramount: platform economy is about the monetization of efficiency, be it transportation, accommodation, advertising, commerce, or any other domain, and data are paramount in achieving maximum efficiency.

More broadly, this is a book about the structuration of life in what Andrejevic and Burdon (2015) call 'sensor society', van Dijck et al. (2018) call 'platform society', or Mau (2019) calls 'metric society'. The premise behind these terms is, nevertheless, the same: the capacity to collect data through '[h]uman-non-human assemblages of sensors' (Chandler 2019: 34), render the world measurable and predictable, and put this newly acquired intelligence to (economic) use (see, notably, Schwab 2017). Moreover, an additional impetus for this analysis is that very soon (if not already) there will simply be no outside of digital platforms as we will either have no alternative but to carry out our activities through them or will be constantly surveyed by them even when not using them (Susskind 2018: 153). Still, this totality should not be surprising: ours is a world which feeds on the 'the technical integration of previously distinct content streams, blending commercially produced entertainment with personalized logistics and everyday speech', which is constantly being synchronised across devices (Athique 2019: 6).

The ubiquity of devices means that our activities and life patterns can be picked up in every detail, and the ease and low cost of both capture and storage means that even the most mundane of bits and pieces are collected (and, likewise, we are encouraged to share them ourselves, e.g. through social media activity or self-tracking). As a result, the mundane has become 'one of the key sites through which Big Data is generated', elevating it from something boring and uninteresting to 'a domain of creativity and improvisation as well as a site of these everyday routines, contingencies and accomplishments' (Pink et al. 2017: 1), all of them being repurposed in a data form for commodification (Zuboff 2015: 79). In this environment of ambient connectivity and data capture, 'we *generate* more than we participate'—again, precisely due to the recording of the most minute and mundane of details—and even on occasions when we *do* actually participate through voluntary sharing activities, this participation also generates data (or, rather, metadata) by itself, independently of *what* is being shared (Andrejevic and Burdon 2015: 20).

As a consequence, populations occupy a dual role in the ecosystem of digital economy: they are simultaneously 'the sources from which data extraction proceeds and the ultimate targets of the utilities such data produce' (Zuboff 2015: 79). In other words, there is an inherent loop, in which humans become double producers of value: of data, which must be treated as capital in itself, not least because it can be directly converted to monetary value through the sale of data to third parties (or through allowing them to use your pool of data), and of direct returns, which could be financial (as in purchasing products or services) or behavioural (such as voting for a candidate). The two kinds of value production are mediated through algorithms that are perhaps best understood as 'sociotechnical systems' that 'link society, technology and nature in a mesh of relations'—in fact, their operation is often *primarily* concerned with *relating* (or folding) things, thereby, again, not only analysing but also actively forging the world (Lee et al. 2019: 2). The preceding might, at first sight, seem to be firmly taking agency away from humans;

however, that is not entirely the case: while agency is, undoubtedly, severely dislocated and made fluid—and the posthumanist framework applied in this book shows that perhaps there was not even much to dislocate in the first place—as humans fold things into algorithms while algorithms fold things elsewhere (and fold humans into things) to equal measure (Lee et al. 2019: 2). As a result, then, this book should be seen as a response to the need to focus on the 'interplay of human capabilities and the capacities of more or less smart machines' (Pentzhold and Bischof 2019: 3). In this context, it is important to note that there is one further element of governance that this book is concerned with—the conditioning of human beings that both encompasses and goes beyond the architectural level. And while Frischmann and Selinger (2018: 6) are perhaps slightly overly dramatic in their assertion that 'We are being conditioned to obey. More precisely, we're being conditioned to want to obey', it is clear that our behaviour has become open to ever more pervasive monitoring while the data collected through such monitoring enables actors leave us unable *not to choose* their preferred options. Overall, then, algorithmic governance can perhaps be best described as polymorphic.

1.1   An Outline of the Book

Following from the broad-stroke picture presented here, the second chapter opens with a description of the collection of data in today's digital environments as well as the nature and use of Big Data, leading to a broader emphasis on datafication and data harvesting. In this context, better—broader and more efficient—data collection is seen as the motivation for and the driving force behind innovation that is often substituted for reality in expectation that data are to give better access to the world. Essentially, this condition is perhaps best described as faith in data, leading to their ever-greater employment in various spheres of life. Moreover, one could also witness a broader cultural shift towards acceptance and, in many cases, furthering of datafication (e.g. self-tracking), leading to agglomerations of bodies and data. This datafication would not be possible without the platform economy, particularly in terms of provoking the generation of and then collecting the data but also with platforms serving as data infrastructures. The pattern is expanded to ever more aspects of everyday life, including the physical environment through the Internet of Things. In a reverse of the classical process whereby the objects of attention would have been identified first and only then data about them would be collected, data themselves are seen to suggest patterns and objects of interest as well as the questions that ought to be asked. All this makes it reasonable to collect as much data as possible pre-emptively. Such collection, in turn, enables pre-emptive actions, i.e. targeting of individuals prior to them making a likely choice (or recommending a choice to them) or to committing an act (as in predictive policing).

Next, the third chapter opens with describing the role of algorithms and (mis)uses of the term in mainstream literature. An initial matter of focus is the use of algorithms in crunching data and providing recommendations, solutions, and decisions derived from the data. Subsequently, attention shifts to the architectural role of algorithms in shaping everyday lives as well as the orchestration of life through determining, ascribing, and prescribing choices, groups, categories, and objects to be made available for interaction. A key role here is played by predictive and interactive machine learning algorithms possessing a noteworthy degree of autonomy. Relating to the previous chapter, such algorithms are demonstrated to be the main structural factors of online platforms, shaping the content and flow of information. As a result, since they embody commercial decisions made by the platform creators, even a minor tweak with a view of increasing profitability and data collection efficiency can have major repercussions in social and political contexts as well as for businesses relying on platforms for their activities. Simultaneously, contrary to some popular depictions, algorithms are to be seen as value-laden, based on the choices that code-writers have made. Hence, the supposed objectivity and neutrality of algorithms only entrenches some values at the expense of others, typically based on commercial, rather than public, interest.

Matters become even more complicated when one considers that machine learning algorithms are used not only to gauge information but also to modulate attention and, even more so, for profiling, for both private and public purposes, indicating the move of algorithms towards the governance of our societies. To that end, algorithmic governance must be conceived as a move from a disciplinary logic to the logic of control. In fact, societies become co-constructed by code that determines what gets to be in the world. This chapter concludes with a discussion of the relative differences and similarities between code and the traditional regulator—law. In addition to opacity, some of the differences highlighted include the malleability of code and, as a result, lack of constancy and predictability of regulation and constant A/B testing, meaning that even members of the same community will be simultaneously affected by different digital regulatory regimes. Hence, with online platforms moving towards becoming operating systems of the whole life, Lessig's famous thesis that 'code is law' necessitates debate.

Meanwhile, Chapter 4 opens with a discussion of personalization and seamless tailoring that stem from algorithmic analysis of data and placement of content. This wrapping of users in custom-made experience cocoons is seen as necessary in an economy based on attention and experience. Moreover, such personalization must extend even beyond the cognitive level—it must also entail emotional and affective flows. Hence, algorithmic governance must be understood not only in terms of architectural affordances but also as extending deeper, manifesting clear affective capacity. That is paramount to success in times of cognitive overload, when attention has become the scarcest commodity. And it is here that algorithms come in handy, helping to determine the stimulus necessary for attracting attention and/or triggering the desired reaction in target audiences. On the other hand, it is demonstrated that satisfaction is an innate physiological need that algorithms can learn to fulfil effectively. Hence, it comes as no surprise that consumer experience and satisfaction are paramount not only in sales but also in (algorithmic) governance, attracting and managing attention flows on both individual and group levels. No less importantly, algorithmically derived knowledge and the capacity to strategically place informational and emotional triggers open up the capacity for nudging individuals towards particular decisions. Hence, a discussion of biases and their outcomes is provided, particularly from a perspective informed by behavioural economics. As a result, attention shifts to the power of choice architects (i.e. those in charge of code) in not only rearranging but effectively creating the reality encountered by users, making it impossible for the latter not to submit to the nudge. The obvious problem here is, of course, that while the decisions may be utility-maximising for the choice architects, the same might not be the case for the users. As a result, a paramount question of agency is opened up.

Picking up on the issue of agency, Chapter 5 shows algorithms, particularly in the context of machine learning, to be capable of increasingly autonomous agency. Capacity to affect others, and not the biological/digital divide, is what matters. Despite algorithms partly retaining their tool-like nature, their world-producing capacities are on the rise. Of course, there is still a fair degree of human shaping (code writing) involved. On the other hand, such shaping is in itself premised on algorithmic feedback loops. At the same time, humans are still important as generators of data essential for algorithms. Simultaneously, though, such data generation happens under algorithmically determined conditions. In fact, fluidity and ambiguity of agency must be seen as key: instead of artificially ascribing it to particular actors, one must concede to it being embedded in human-digital agglomerations. Hence, agency becomes distributed and, in part at least, devoid of causal chains. Consequently, the human is de-privileged, rendered just one of many objects with some agentic capacities. Meanwhile, the self is accordingly fragmented and dispersed, with multiple datafied effigies of ourselves (or parts of ourselves) residing on multiple databases, activated depending on choice, convenience, relevance, availability, etc., thus further adding to the fluidity and malleability of the new agentic environment.

The above definitely calls any anthropocentric presumptions into question. As a result, agency embedded in human-digital agglomerations is better understood from a posthumanist framework. While there are numerous different strands of posthumanism, they are united by an understanding of human beings as merely some of the pieces of a broader context and as significantly malleable, thereby questioning the very premises of being human. Plasticity and malleability, embeddedness, and shared agency clearly denote today's human-digital agglomerations, blurring lines and distinctions.

Subsequently, Chapter 6 contrasts the agency embedded in human-digital agglomerations with more traditional accounts of regulation, particularly from rights- and legitimacy-based perspectives. For example, the issues around disassembling of the digital datafied effigies and whether we are entitled to a digital equivalent of the inviolability of the human person constitute one of the key clusters of ideas animating this chapter. Another noteworthy matter is the importance of technical rules governing the collection and use of data. Hence, rule-making, through both code and technical specifications, is the primary means of governance that spans the domestic, the private, and the public domains. Again, from a rights-based perspective, the necessity for humans to know the law one is subjected to, particularly when being charged with wrongdoing, is undermined by the malleability and the opacity of algorithmic rules. In a similar manner, predictive algorithms used for governance have an uneasy relationship with presumption of innocence: proactive limitation just because one is deemed likely (or is algorithmically associated with a group likely) to commit a wrongful act is, in effect, punished in advance through putting stricter limits on actions. Moreover, since even regulations issued by governments are becoming impossible to enact in practice without cooperation of the major technology companies, the latter effectively become embedded in legal and political systems, leading to hybridisation of governance. The chapter then moves to exploring legitimacy issues, probing whether it is at all possible to speak of legitimacy in an environment where algorithmic tailoring wraps individuals in experience cocoons whereby the collective level becomes ever more difficult to conceive. As a corollary, one must also question to whom the wielding of power is to be justified when communities become dis-imagined and algorithmically reconstituted. Certainly, algorithmic analysis of data and the capacity to capture audience sentiments can have positive legitimacy effects as well. Primarily, that pertains to increased responsiveness to electorate demands, hopes, fears, and aspirations, which is a clear benefit as far as the teleological, results-oriented approaches to legitimacy are concerned. However, it is not far-fetched to argue that in this case the entire electoral process becomes void, losing the substantive content of legitimacy.

REFERENCES

Andrejevic, M., & Burdon, M. (2015). Defining the Sensor Society. *Television & New Media, 16*(1), 19–36.

Athique, A. (2019). Integrated Commodities in the Digital Economy. *Media, Culture and Society.* https://doi.org/10.1177/0163443719861815.

Chandler, D. (2019). Digital Governance in the Anthropocene: The Rise of the Correlational Machine. In D. Chandler & C. Fuchs (Eds.), *Digital Objects, Digital Subjects: Interdisciplinary Perspectives on Capitalism, Labour and Politics in the Age of Big Data* (pp. 23–42). London: University of Westminster Press.

Cowley, R. (2019). Posthumanism as a Spectrum: Reflections on Paul Rekret's Chapter. In D. Chandler & C. Fuchs (Eds.), *Digital Objects, Digital Subjects: Interdisciplinary Perspectives on Capitalism, Labour and Politics in the Age of Big Data* (pp. 95–100). London: University of Westminster Press.

Frischmann, B., & Selinger, E. (2018). *Re-Engineering Humanity.* Cambridge and New York: Cambridge University Press.

Fuchs, C., & Chandler, D. (2019). Introduction. In D. Chandler & C. Fuchs (Eds.), *Digital Objects, Digital Subjects: Interdisciplinary Perspectives on Capitalism, Labour and Politics in the Age of Big Data* (pp. 1–20). London: University of Westminster Press.

Hill, D. W. (2019). The Injuries of Platform Logistics. *Media, Culture and Society.* https://doi.org/10.1177/0163443719861840.

Lee, F., et al. (2019). Algorithms as Folding: Reframing the Analytical Focus. *Big Data & Society.* https://doi.org/10.1177/2053951719863819.

Mau, S. (2019). *The Metric Society: On the Quantification of the Social.* Cambridge and Medford: Polity Press.

McAffee, A., & Brynjolfsson, E. (2017). *Machine, Platform, Crowd: Harnessing Our Digital Future.* New York and London: W. W. Norton.

Pentzhold, C., & Bischof, A. (2019). Making Affordances Real: Socio-Materia Prefiguration, Performed Agency, and Coordinated Activities in Human-Robot Communication. *Social Media + Society.* https://doi.org/10.1177/2056305119865472.

Pink, S., et al. (2017). Mundane Data: The Routines, Contingencies and Accomplishments of Digital Living. *Big Data & Society.* https://doi.org/10.1177/2053951717700924.

Schwab, K. (2017). *The Fourth Industrial Revolution.* London: Portfolio.

Srnicek, N. (2017). *Platform Capitalism.* Cambridge and Malden: Polity Press.

Susskind, J. (2018). *Future Politics: Living Together in a World Transformed by Tech.* Oxford and New York: Oxford University Press.

van Dijck, J., Poell, T., & de Waal, M. (2018). *The Platform Society: Public Values in a Connective World.* Oxford and New York: Oxford University Press.

Zuboff, S. (2015). Big Other: Surveillance Capitalism and the Prospects of an Information Civilization. *Journal of Information Technology, 30,* 75–89.

CHAPTER 2

Data: The Premise of New Governance

Abstract The collection, analysis, and use of data are the defining features of today's world. In fact, it is safe to say that today's world has been fundamentally datafied, particularly courtesy to the proliferation of platforms that provide structure to (and, in fact, drive) contemporary social and business practices. No less importantly, this process has encompassed not only public life but also the human body and private environments. To that effect, the observability and predictability of individuals and the capacity to turn their lives into valuable commodities have become highly pervasive, and the more so the more the dominant platforms accumulate and strengthen their network effects. As a result, then, the position of the human person is profoundly altered in the world where there is no more outside to commercial(ised) data.

Keywords Platformisation · Datafication · Internet of Things · Self-tracking · Data infrastructures · Prediction

It is by no means an exaggeration to contend that today's economy has become 'centred upon extracting and using a particular type of raw material: data' (Srnicek 2017: 39). At its most basic, big data can be described in terms of three Vs: 'the extreme volume of data, the variety of the data types, and the velocity at which data must be processed' (Kelleher and Tierney 2018: 9). This data-driven economy has, however, also branched out to other areas of human life, ultimately affecting the

© The Author(s) 2019
I. Kalpokas, *Algorithmic Governance*,
https://doi.org/10.1007/978-3-030-31922-9_2

nature of human life and human existence as such. Of prime importance, therefore, are novel and creative ways of data extraction, ranging from platformisation to tracking of online practices to (self-)tracking through wearable devices of various sorts to the Internet of Things. Data harvesting is at the heart of these innovations and, through their ever-growing role in everyday lives, pervades everything we do, in the most mundane of details. Such cornucopia of data are employed to tailor services and information and predict human action even prior to the action taking place—an enticing possibility for both corporate actors and states—in exchange for convenience, consumer satisfaction, and (alleged) optimisation of the self and one's own surroundings.

2.1   DATA DATA EVERYWHERE...

Today's world is characterised by our ability to collect and analyse unprecedented amounts of data—in fact, it is often not impossible to 'process all of it relating to a particular phenomenon' (Mayer-Schönberger and Cukier 2017: 12). Indeed, big data and their analysis only properly begin when both the amount of data and the complexity of patterns therein become impossible to deal with manually (Kelleher and Tierney 2018: 4)—otherwise, data can still be considered 'small'. That, in turn, has transformed the ways in which the world is known. Indeed, traditionally, when dealing with large numbers (e.g. surveying a population) one had to rely on samples that, although representative of the whole, were incapable of revealing details of a more granular nature. Meanwhile, the employment of big data enables the discovery of details, providing 'an especially clear view of the granular: subcategories and submarkets that samples can't assess' (Mayer-Schönberger and Cukier 2017: 12–13). Hence, understanding of groups, processes, etc. as well as their predictability is greatly enhanced. Moreover, in some cases at least, data-based solutions can even be offered to problems we did not know existed, potentially even prior to us knowing that they existed (Beer 2019: 14).

Still, there is a necessary precondition for such revelation: the world first has to be datafied. Datafication here 'refers to taking information about all things under the sun – including ones we never used to think of as information at all [...] and transforming it into a data format to make it quantified', thereby unlocking 'the implicit, latent value of the information' (Mayer-Schönberger and Cukier 2017: 15). In other words, a new—digital—layer of exploitation is added onto the world, making the entirety of it one big generator of data. And as new uses of data are come up with, they are typically sold in the form of claims about the improvement and optimisation of everything from one's personal lifestyle to the economy and the environment, thereby positing data as a miracle potion that can answer our dreams (Beer 2019: 14). And while such representations are not always necessarily erroneous and/or misleading, there are many important practicalities to be considered.

At the heart of why big data are so appealing lies the promise of superseding theories and models with 'the materiality of the world', supposedly providing an 'unmediated access' to the world-as-it-is, in all its multiplicity and complexity, as well as to 'real' abstraction-free individuals inhabiting it—in other words, a 'better access to reality' (Chandler 2015: 846–848). It is, therefore, no surprise that, as Sadowski (2019: 1) claims, industries across a variety of domains from technology and manufacturing to finance, consulting, and others have come to treat data as an important form of capital. Examples may include 'emails, videos, audios, images, click streams, logs, posts, search queries, health records, and more' (Kemper and Kolkman 2018: 1) as well as 'trace data', i.e. metadata that we reveal while browsing, sharing, conversing, and otherwise going about our everyday lives (Faraj et al. 2018: 64; Kelleher and Tierney 2018: 199). After all, as stressed by Caplan and boyd (2018: 4), 'in a world where surveillance is the norm, merely existing in the world means you are structured into the technologies and systems that structure most of social life today'. Essentially, today there is little chance of *not* living a life of a data subject who is simultaneously a generator of data and somebody at the receiving end of data analysis practices.

Moreover, it is vital to understand that the line between provision of an agreement to surrender one's data and lack thereof is becoming progressively blurred. While so-called 'captured data', i.e. the result of direct and explicit observations or measurements, at least require voluntary disclosure, the more prevalent form, so-called 'exhaust data', are merely 'a by-product of a process whose primary purpose is something other than data capture', such as retail or social networking, and therefore are simply collected by default (Kelleher and Tierney 2018: 52). Undoubtedly, the latter data stream is worrying due to the absence of user control. However, it is also worth stressing that the capacity to uncover unanticipated patterns in data as well as the repurposing of data for ever new uses means that even the data that had been surrendered voluntarily can have unpredictable consequences and be put to unpredictable, previously unanticipated uses, rendering any prior consumer agreement useless (Kelleher and Tierney 2018: 200).

Indeed, as the recording and storage of data is cheap (and is constantly getting even cheaper), and data can be captured without much effort (of both the collectors and the data subjects), we have reached a situation in which, absent legal regulation, 'it is easier to justify keeping data than discarding it' (Mayer-Schönberger and Cukier 2017: 101). That is one further reason that animates the shift 'from collecting some data to gathering as much as possible, and if feasible, getting everything: n=all' (Mayer-Schönberger and Cukier 2017: 28). And as data amass, so does the pressure to do something with them—after all, the sheer amount of data can be seen as potentially creating opportunities for new and creative uses (Beer 2019: 4). In this situation, all data are progressively deemed to be valuable as such and in advance simply because they exist—after all, 'most innovative secondary uses haven't been imagined when the data is first collected' (Mayer-Schönberger and Cukier 2017: 100, 153). As a result, collecting as much data as possible, at least due to a fear of missing out on some bits that are to ultimately turn important and valuable, is a natural default strategy.

Still, the relevant benefits are not derived from data directly—a crucial step is extracting patterns and other types of information that are 'both nonobvious and useful', helping us address a particular problem (Kelleher and Tierney 2018: 4–5). In other words, the requirement here is for the production of 'actionable insights', the term referring to precisely what it says on the tin: These are insights because they give us a glimpse into something nonobvious that has a practical use that can be acted upon (Kelleher and Tierney 2018: 5). This extraction of patterns and other meanings is a key challenge that data analysis must overcome because most of the available data tend to come in an unstructured form, manifesting diverse internal structures and characteristics (Kelleher and Tierney 2018: 48–49; see also Mayer-Schönberger and Cukier 2017: 13–14). Moreover, such data are typically *raw*, i.e. a cornucopia of singular attributes, measurements or traits from which value must be derived by applying some function (Kelleher and Tierney 2018: 49–51). Certainly, such analysis is not something that humans are capable of by themselves. Instead, algorithms are tasked with ploughing through datasets of 'immense quantity and breadth', often producing insights that are 'counterintuitive, but uncannily accurate', providing data-rich actors with the capacity of targeting individuals in tailored and customised ways that 'leverage aspects of personality, political leanings, and affective proclivities' (Faraj et al. 2018: 64). As a result, analysis of big data can have not only analytical and prognosticatory but also manipulative use.

Moreover, the sheer abundance of data, when coupled with the falling cost of analysis, results in increasing automation of decision-making processes in both public and private sectors that, in turn, drive the glut for ever more data (Kemper and Kolkman 2018: 1). In fact, it could be easily claimed that '[w]ith most objects being tagged, locations identified, people's attributes marked, behaviour traced, and interactions mapped, all aspects of working and living can be digitally represented and quantified' (Faraj et al. 2018: 64) and predictions made regarding future actions, thoughts, and likes courtesy of past and present behaviour of the target groups and individuals (Newell and Marabelli 2015: 4). And, because every set of data that can be composed is likely to have value, albeit one not necessarily yet known, it is only natural that everything that is up for grabs is captured, signalling the advent of data as *the* key resource, succeeding factories and intellectual property that had in their own times been the most valuable properties and foundations of business models (Mayer-Schönberger and Cukier 2017: 15–16).

From the above, it is not surprising that Beer (2019: 14) even refers to a *faith* in data whereby the latter is thought and expected to provide all the answers; solution to any problem would therefore be deceptively simple—collect more data, because if only we had enough of them, solutions would come by almost miraculously. Similarly, expectations and promises abound with a view that 'a data-driven approach to governance is the best way to address complex social problems, such as crime, poverty, poor education, and poor public health', if we only manage to instal enough sensors to monitor every aspect of our society, merge the data thus collected, and devise algorithms for their analysis (Kelleher and Tierney 2018: 196). That, in turn, leads to two major problems. The first of them is the increasingly technocratic nature that the society assumes, with ever more areas of life succumbing to impersonal data-driven regulation; the second, meanwhile, is the so-called 'data creep', or repurposing of data and data collection tools in order to regulate in previously unintended ways and unlock previously unintended insights (Kelleher and Tierney 2018: 196–197; see also Frischmann and Selinger 2018). In both cases, the promise of data becomes a legitimising tool for ever-deeper encroachment on everyday lives of individuals.

There has been, in effect, a cultural shift leading towards the acceptance of and, perhaps, even the desirability of quantification of human beings and their activities (Papsdorf 2015: 995–996), enacted by both humans themselves (self-tracking) and corporate and state actors (data collection and surveillance). That accumulation has ultimately led to '[t]he creation of pools of data that cover a variety of individuals, extended periods of time, and diverse subjects' (Papsdorf 2015: 997). In part, that is due to the social nature of our lives (the necessity to interact or, even more explicitly, to share), but datafication has also been extended to domains that had previously been private par excellence (such as bodily functions). In effect, we have become '"leaky bodies" in relation to data', such bodies being themselves in 'continual transformation' through data feedback loops, ultimately condescending to 'meshworks of entangled lines of body, data and technology' (Tucker 2018: 40). As will be argued later in this book, the latter are even becoming less and less separable, being agglomerated into relational assemblages.

While automation of collection and analysis is necessary in order to make full use of today's developments, such interactive agglomerations of bodies and data are transformative in themselves, shedding new light on the standing of human beings and paving the way for posthuman transformations that will be conceptualised further in the book. At the very least, such a transformation relates to the way in which interactions of humans and data reposition (in favour of data or, rather, data-crunching algorithms) the ways in which interactions, content, and underlying conditions of everyday life are formed, directed, and experienced (Murray and Flyverbom 2018: 7). In other words, with big data comes big power, although it is less clear about responsibility.

It might, to some extent at least, be tempting to think about this datafied environment as a Panopticon with a potentially disciplinary function—Foucault (2012) would be an obvious reference here. However, reliance on either Bentham's original idea of the Panopticon or Foucault's appropriation of it (or the broader disciplinary framework that Foucault develops) would be misleading because in the age of data, quasi-disciplinary aims are achieved with alternative means. While the original idea included disciplining behaviour through the *known* potential of being observed or imposition of norms and regulations that are known to be such, with big data, always being seen is no longer potential—it is reality. Moreover, while in the Panopticon the subject being disciplined had to be aware of a potential observer or the regulations to be observed, the key today is for the subject to *forget* about being observed. Indeed, we are even encouraged to reveal ourselves inasmuch as possible, only for the revealed attributes and traits to be used to gently and pleasurably nudge us in predefined directions (as shown in the chapters that follow). Straightforward discipline would be hard to sell to a consumerist society. The goal, instead, is to retain consumer satisfaction with themselves being nudged around while preserving the impression of unhindered human agency.

2.2   Immersion and Generation

Perhaps the most common occurrence of data collection is that through platforms of the most varied sorts (social media, ride sharing, holiday booking, dating, etc.), connecting users not only to friends, acquaintances, and the content shared by them but also with potential dates, ride sharers, service providers, holiday hosts, etc. while intermediating the flows of data along the way (Beer 2019: 3; see also McAffee and Brynjolfsson 2017). In effect, then, 'platforms are digital infrastructures that enable two or more groups to interact', arrogating the role of 'intermediaries that bring together different users: customers, advertisers, service providers, producers, suppliers, and even physical objects' (Srnicek 2017: 43). However, platforms themselves are clearly more than just specific interfaces for interaction—they themselves become data infrastructures due to the integration of data held (Langlois and Elmer 2018: 5; see also Nieborg and Helmond 2018: 4), enabling all kinds of users, such as political, business or other actors, information or entertainment providers, etc., to provide their own offering to the public (as in social media-centric political campaigns). Simultaneously, however, such actors are rendered dependent upon the infrastructural layer.

Effectively, platforms have become 'an efficient way to monopolise, extract, analyse, and use the increasingly large amounts of data that were being recorded' (Srnicek 2017: 43) courtesy of their 'network effects', whereby 'the more numerous the users who use the platform, the more valuable that platform becomes for everyone else' due to its ever-expanding data harvesting capacities (Srnicek 2017: 45). In other words, the more individuals are using the platform the more data they leave and the more tailored the services can become; simultaneously, the more providers are using the platform, the greater the variety of services that can be offered, thus enticing more users; and the more users and providers are using the platform, the more interactions take place, therefore, even more data can be extracted; all that attracts even more users and service providers and so on. As that happens, the lack of both customers and service providers outside the platform further increases the costs of not being in, exerting pressure to join on even the most resistant. Moreover, competitors to today's dominant platforms are effectively barred from entry because of the data that the already dominant players are able to extract from their users and incorporate back into their architecture, thus offering a product that simply better suits audience needs (Mayer-Schönberger and Cukier 2017: 115). Hence, the drive towards monopoly of both market and data, inherent in platformisation, is laid bare.

Furthermore, even though platforms (particularly those for social interaction and search) still tend to be the most widely discussed tools for the collection and use of data, nowadays one can observe expansion of the practice across a broad spectrum of apps as well as the Internet of Things, ubiquitous computing, and smart cities, thus interconnecting diverse actors and layers of infrastructure (Iliadis 2018: 219). In fact, datafication has by now intruded the human body itself, courtesy of wearable devices that are 'hinges between the body and the network: ways of raising the body's own processes directly to the network, where they can be stored or mined for insight like any other data set' (Greenfield 2018: 33), thereby establishing a direct body-data link. Crucially, self-tracking through wearable devices 'is becoming ordinary', likely because of the completely mundane, seemingly unexceptional nature of the activities being tracked, such as 'the food we eat, our sleep, how much we exercise, the rhythm of the heart and the work we perform' (Lomborg et al. 2018: 4591). The latter is a clear example of the stealth of datafication: because much of the data appear so boring, so seemingly irrelevant to anyone but oneself (or even to the subject surrendering the data), so deceptively inapplicable in contexts outside their collection that they are easy to surrender without giving much thought.

Moreover, datafication and (self-)subjection to data becomes easier to swallow if it is rendered as a trend, an emergent cultural phenomenon that points to greater, better, or at least more exciting ways of doing things. Returning to the case of self-tracking, an entire ideological-cultural apparatus seems to be developing in connection with such devices, organised around the premise of 'self-knowledge through numbers', i.e. quantification of 'as many aspects of your life as possible' with the aim of using such data 'to change, and improve, your life' (Didžiokaitė et al. 2017: 1471). The key promises here are those of '*transparency* and *self-optimisation*', maximisation of health and work performance, as well as control and mastery of the self, ultimately making oneself 'better' (Didžiokaitė et al. 2017: 1473–1475; see also Beer 2019: 4). Extreme datafication thereby becomes seen as a tool for greater liberation which, unlike other liberations, becomes a convenience rather than a struggle or a chore: as Langlois and Elmer (2018: 10) pinpoint, the tracking of many aforementioned parameters 'does not require any kind of attention from the subject, and requires a preconscious routine habit of keeping one's smart watch on'. As a result, datafication becomes easier to 'sell' to the consumers and, therefore, easily takes ever more pervasive forms.

While at face value it might seem that self-tracking is something merely concerned with the 'self', aiding self-awareness and self-improvement, in practice it is fundamentally communicative: when tracking our lives (not just bodies but also habits, activities, relationships, etc.), 'we communicate with a system, with ourselves, and the social world', thereby ultimately entering an imperceptible blend 'where both users and their apparatuses integrate with, and contribute to, socio-technical flows on multiple levels' (Lomborg et al. 2018: 4591). At the same time, potential opens up for datafied control that is 'ambient' as 'decisions once thought to be purely personal – sleep cycles, nutritional patterns, exercise habits' can increasingly be subjected to monitoring, control, and intervention (Greenfield 2018: 197). This time, a reference to disciplinary regimes a la Foucault might seem more meaningful. What has to be kept in mind, nevertheless, is that the data-based ascription of habits, behaviours, and routines is merely a by-product and a tool for further extension of datafication and not an ordering instrument on its own (that having been said, algorithmic application of data will subsequently be demonstrated to have strong ordering proclivities and governance both *can* and *is* built on the scaffolding of datafication).

In addition, data collection does not merely *intrude* our bodies: it *surrounds* them as well. This is the context in which the so-called Internet of Things should be understood: 'as a platform-driven extension of data recording into everyday activities', enabling the full spectrum of our daily lives to be recorded in the most mundane of details (Srnicek 2017: 100) In that sense, the very term 'Internet of Things' is imprecise—it is not about things but about humans or, even more narrowly, their data. (Vaidhyanathan 2018: 99). What dominates in this context is a pre-emptive capture logic whereby, as service provider, one is incentivised to 'trawl up everything you can' for the reason that it is impossible to foresee what value (if any) will be derived from a particular data unit at a particular time in the future (Greenfield 2018: 41). In other words, the more we invest in smart connected devices, the more we immerse ourselves into the data pond, in many ways rendering ourselves dependent on data flows as ever more devices are adopted and entrusted with everyday functions. Simultaneously, we become further incentivised to contribute to data generation, normalising the datafication of life in exchange for convenience.

Still, actors collect data not merely because they can and because there are plenty available but also because any use to which they are going to be put is going to itself be data-intensive. As McQuillan (2016: 2–3) stresses, in order to predict somebody's purchase or any other likely action, one must take into account a cornucopia of correlated features, such as 'a browsing history of hundreds of urls, the browser used, location of the user, time of day, weather conditions, their friendship networks on social media, and so on'. But even that is by no means a full picture, since the very capacity to make such a prediction is merely the tip of an iceberg, the end result of a massive machine learning process for the data-crunching algorithm—in all likelihood, that learning process will have necessitated 'millions of training data points from other people's previous purchases' in order to tease out the recurrent mechanisms and patterns pertaining to human choice (McQuillan 2016: 2–3). Hence, the data glut appears to be self-reinforcing: Data reveal to us the need for more data, and to analyse that one needs even more data, and so on: Whatever amount of data we have is merely an impetus for striving for more.

Likewise, data promise greater conveniences: If our devices know ever more about ourselves, then it means that they can also offer services that are better tailored for our predispositions, likes, and interests. But such conveniences, in turn, necessitate even more data. That unavoidably leads to concentration within the digital environment: 'the more activities a firm has access to, the more data it can extract and the more value it can generate from those data and therefore the more activities it can gain access to', transforming key digital players into agglomerations that span acquisitions across most areas of human life in order to centralise data and offer more predictive, intuitive, and seamlessly tailored services in order to lure in even more users and, therefore, even more data (Srnicek 2017: 95). The users of platforms and other services thereby become 'commodities at least as much as they are consumers', their attention (or, at least, information about how that attention can be captured) being either harvested for the platform owners' direct service provision purposes or sold and/or bought to and from third parties or both (Sunstein 2018: 229). Indeed, as more and more conveniences are introduced, from scheduling, recommendation, or purchasing services to personal assistants and as they seem to be working more and more seamlessly—indeed, intuitively—the higher a price we pay in our data (Tiku 2018).

In a broader context, such collection is part and parcel to the broader reverse of the classical toolkit that relied on first identifying the persons, trends, and other objects/phenomena of interest and then collecting all available information about them—the current modus operandi involves aggregation of data *prior* to determining its use and potential in order to intervene even *prior* to the (un)desired actions, developments, or occurrences with a view to either reinforce or pre-empt the relevant causal factors (Lyon 2014: 4). The more data are or become available, the more it becomes possible to identify 'features and useful classifiers' that are discerned in a progressively robust manner' (Greenfield 2018: 221–222). Ultimately, as the data are 'iteratively resolved in ever-greater fidelity, the patterns themselves begin to suggest the questions that might be asked of them' (Greenfield 2018: 211). Once again, the predictive and pre-emptive nature of data analysis becomes clear. However, particularly in case of pre-emption of undesirable consequences, a prominent question arises: to what extent we should be allowed to act upon such information (e.g. applying pre-emptive measures against an individual prior to them committing a crime) and to what extent we can be certain that the patterns and, therefore, the predictions are correct.

Crucially, the reliability of data and their analytics is not unquestionable. Such scepticism primarily relates to the very origin of big data—instead of being collected purposefully and systematically in accordance to predefined categories and under strictly controlled conditions (as customary with traditional data collection methods), big data typically simply 'happen' and are collected through the unfolding of everyday life simply because this or that activity happens to be tracked by a particular actor (Müller et al. 2016: 289–290). As Greenfield (2018: 210) puts it, '[w]henever we say "data", [...] what we're really referring to is that subset of the world's infinite aspects that have been captured by some instrument or process of measurement'. Hence, data are neither objective nor neutral but merely the result of practices immanent to the very process of collection. Contrary to the promise of greater transparency and heterogeneity, whereby datafication was supposed to shine light on ever more everyday practices on the micro level and thus emancipate a plethora of previously underrepresented voices in the process (Chandler 2015: 850), the actual reality appears to be one of haphazard and almost random datafication, a process in which the nature of the collecting tool affects the nature of what is collected.

Finally, distinction has to be made between data, i.e. 'information that something happened' on the one hand and knowledge, or 'information about why something happened' on the other: While a given set of data might also involve pieces of knowledge, such a presence 'is not a necessary condition' (Srnicek 2017: 39). In an even more granular fashion, it is widely acknowledged that there is a continuum of 'data, information, knowledge and wisdom', with significant transformations necessary in order to proceed from one node to another: the world must be measured for the purpose of data production, the latter must be organised for the purpose of information production, which subsequently must be synthesised with other elements, such as experience, for knowledge production, and only then, in an (as yet) unknown way, wisdom is arrived at (Greenfield 2018: 210). As a result, mere possession of big data, which are typically messy, of varying quality, and often residing on multiple platforms and servers (Alpaydin 2016: 154), in itself implies neither information, nor knowledge but merely the potential for those.

In the above context, it is not a coincidence that Srnicek (2017: 40) likens data to oil: in addition to their key role in today's economy, '[j]ust like oil, data are a material to be extracted, refined, and used in a variety of ways'. In a similar sense, Braidotti (2013: 62) refers to 'life-mining' and the 'visibility, predictability and exportability' of the results: Life itself contains valuable resources that have to be extracted and exported for refinement, thereby unlocking hidden wealth and possibilities. On the other hand, it is not entirely without reason that other authors are somewhat critical of such an approach (see Sadowski 2019). For them, the term 'data mining' is misleading, merely reinforcing 'regimes of data accumulation' by a 'framing of data as a natural resource that is everywhere and free for the taking' and should therefore be substituted with 'data manufacturing' to better reflect the work that goes into the acquisition and making sense of data (Sadowski 2019: 2). Nevertheless, the two frames are less in opposition than it might seem: Even Srnicek's or Braidotti's data-as-resource approach is also illustrative in its emphasis on things that have to be done *to* data before one can proceed to doing things *with* data.

Crucially, it is primarily 'data that have been *worked on*' that have commercial viability and are practically significant (Srnicek 2017: 56–57; see also Pasquale 2015: 21). In a typical model, data have to first be collected from a variety of sources, then aggregated, preprocessed, and explored for potential uses; it is only subsequently that machine learning techniques can be applied to extract the value in terms of information that can be used to support and inform decision-making processes (Kelleher and Tierney 2018: 57). In other words, '[a]lthough the talk is often of the data, it is actually the analytics that are seen to be powerful in their envisaged potential': effectively, data only 'come to life and begin to have consequences when they are analysed and when those analyses are integrated into social, governmental and organisational structures' (Beer 2019: 14–15). The preceding has two important implications: first, raw data lacks meaning; second, data, due to their abundance and capture-ability are by now taken for granted, so the key is in doing something meaningful to and with them. And this extra work, necessary for the refinement and analysis of data, due to their sheer volume, is typically performed by algorithms.

REFERENCES

Alpaydin, E. (2016). *Machine Learning: The New AI*. Cambridge, MA and London: The MIT Press.

Beer, D. (2019). *The Data Gaze: Capitalism, Power and Perception*. Los Angeles and London: Sage.

Braidotti, R. (2013). *The Posthuman*. Cambridge and Malden: Polity Press.

Caplan, R., & boyd, d. (2018). Isomorphism Through Algorithms: Institutional Dependencies in the Case of Facebook. *Big Data & Society*. https://doi.org/10.1177/2053951718757253.

Chandler, D. (2015). A World Without Causation: Big Data and the Coming of Age of Posthumanism. *Millennium: Journal of International Studies, 43*(3), 833–851.

Didžiokaitė, G., Saukko, P., & Greiffenhagen, C. (2017). The Mundane Experience of Everyday Calorie Trackers: Beyond the Metaphor of Quantified Self. *New Media & Society, 20*(4), 1470–1487.

Faraj, S., Pachidi, S., & Sayegh, K. (2018). Working and Organizing in the Age of the Learning Algorithm. *Information and Organization, 28,* 62–70.

Foucault, M. (2012). *Discipline and Punish: The Birth of the Prison.* New York: Vintage Books.

Frischmann, B., & Selinger, E. (2018). *Re-Engineering Humanity.* Cambridge and New York: Cambridge University Press.

Greenfield, A. (2018). *Radical Technologies: Thee Design of Everyday Life.* London and New York: Verso.

Iliadis, A. (2018). Algorithms, Ontology, and Social Progress. *Global Media and Communication, 14*(2), 219–230.

Kelleher, J. D., & Tierney, B. (2018). *Data Science.* Cambridge, MA and London: The MIT Press.

Kemper, J., & Kolkman, D. (2018). Transparent to Whom? No Algorithmic Accountability Without a Critical Audience. *Information, Communication & Society.* https://doi.org/10.1080/1369118x.2018.1477967.

Langlois, G., & Elmer, G. (2018). Impersonal Subjectivation from Platforms to Infrastructures. *Media, Culture and Society.* https://doi.org/10.1177/0163443718818374.

Lomborg, S., Thylstrup, N. B., & Schwartz, J. (2018). The Temporal Flows of Self-Tracking: Checking in, Moving on. *Staying Hooked. New Media & Society, 20*(12), 4590–4607.

Lyon, D. (2014). Surveillance, Snowden, and Big Data: Capacities, Consequences, Critique. *Big Data & Society.* https://doi.org/10.1177/2053951714541861.

Mayer-Schönberger, V., & Cukier, K. (2017). *Big Data: Th Essential Guide to Work, Life and Learning in the Age of Insight.* London: John Murray.

McAffee, A., & Brynjolfsson, E. (2017). *Machine, Platform, Crowd: Harnessing Our Digital Future.* New York and London: W. W. Norton.

McQuillan, D. (2016). Algorithmic Paranoia and the Convivial Alternative. *Big Data & Society.* https://doi.org/10.1177/2053951716671340.

Müller, O., et al. (2016). Utilizing Big Data Analytics for Information Systems Research: Challenges, Promises and Guidelines. *European Journal of Information Systems, 25,* 289–302.

Murray, J., & Flyverbom, M. (2018). Datastructuring: Organizing and Curating Digital Traces into Action. *Big Data & Society.* https://doi.org/10.1177/2053951718799114.

Newell, S., & Marabelli, M. (2015). Strategic Opportunities (and Challenges) of Algorithmic Decision-Making: A Call for Action on the Long-Term Societal Effects of 'Datification'. *Journal of Strategic Information Systems, 24,* 3–14.

Nieborg, D. B., & Helmond, A. (2018). The Political Economy of Facebook's Platformization in the Mobile Ecosystem: Facebook Messenger as a Platform Instance. *Media, Culture and Society.* https://doi.org/10.1177/0163443718818384.

Papsdorf, C. (2015). How the Internet Automates Communication. *Information, Communication & Society, 18*(9), 991–1005.

Pasquale, F. (2015). *The Black Box Society: The Secret Algorithms That Control Money and Information.* Cambridge, MA and London: Harvard University Press.

Sadowski, J. (2019). When Data Is Capital: Datafication, Accumulation, and Extraction. *Big Data & Society.* https://doi.org/10.1177/2053951718820549.

Srnicek, N. (2017). *Platform Capitalism.* Cambridge and Malden: Polity Press.

Sunstein, C. R. (2018). *#Republic.* Princeton and Oxford: Princeton University Press.

Tiku, N. (2018, August 5). The Price of Google's New Conveniences? Your Data. *Wired.* Available at https://www.wired.com/story/the-price-of-googles-new-conveniences-your-data. Accessed 29 Nov 2018.

Tucker, I. (2018). Digitally Mediated Emotion: Simondon, Affectivity and Individuation. In T. D. Sampson, S. Maddison, & D. Ellis (Eds.), *Affect and Social Media: Emotion, Mediation, Anxiety and Contagion* (pp. 35–41). London and Lanham: Rowman & Littlefield.

Vaidhyanathan, S. (2018). *Anti-Social Media: How Facebook Disconnects Us and Undermines Democracy.* Oxford and New York: Oxford University Press.

CHAPTER 3

The Code That Is Law

Abstract The abundance of data in today's world implies the need for algorithms as tools for sorting, ranking, retrieval, interpretation, and decision-making. As a result, algorithms become the moving and driving forces behind today's life, undergirding private, public, and business environments. As a consequence, algorithms acquire an unparalleled power of governance: As they determine the architecture of everyday life, decisions made and actions performed must conform to digitally coded affordances. Hence, the regulatory function of algorithms is explored in this chapter, particularly through comparing and contrasting them to the traditional regulator—law. This analysis reveals important differences, particularly in relation to power and pervasiveness, opacity, interests served, and the general *modus operandi*. As a result, algorithmic governance is seen as a new and distinct form of governance.

Keywords Algorithm · Machine learning · Architecture · Governance · Law · Decision-making

Data, discussed in the previous chapter, can certainly wield power. They are, however, not particularly valuable in their raw form and need to be refined and analysed. It is here that algorithms—or computer code more broadly—enter the fore as tools for not only extracting valuable information from data but also acting upon that information, weighing and analysing even seemingly disparate and disjointed clues in order

© The Author(s) 2019　　　　　　　　　　　　　　　　　　　　27
I. Kalpokas, *Algorithmic Governance*,
https://doi.org/10.1007/978-3-030-31922-9_3

to deliver an answer to a query, an insight into ongoing processes, or a prediction about the future. To that effect, algorithms are to be seen as exerting an ever-greater influence on everyday lives, determining the informational, social, and material living conditions as well as enabling and disabling practices for both individuals and businesses. As a result, algorithmic governance, not dissimilar to that through law (although, arguably, more pervasive) is a naturally ensuing condition. And yet, the inherent non-transparency of algorithms makes this form of governance both less perceptible and less accountable. The purpose of this chapter is, therefore, to critically examine this infrastructure of today's life in order to tease out the critical junctions that mark the shift towards posthuman law.

3.1   THE ROLE OF THE ALGORITHM

Algorithm, by definition, 'includes any rules that humans and/or computers can follow' (Lee 2018: 3), being 'a sequence of instructions that are carried out to transform an input into an output' (Alpaydin 2016: 16). Essentially, therefore, the term typically denotes 'an abstract, formalized description of a computational procedure' (Dourish 2016: 3), which 'autonomously makes decisions based on statistical models or decision rules without explicit human intervention' (Lee 2018: 3). An attempt at a definition, however, is notably obscured by the broad scope of their use and, thus, complexity: as Janssen and Kuk (2016): 372) note, algorithms can equally carry out 'simple calculations or highly complex reasoning tasks'. Moreover, algorithms do different things for different people: whereas for the programmer they embody a command structure, for the user they are primarily assistants, e.g. retrievers, rankers, and organisers of content or search results (Bucher 2018: 49). As such, they are 'problem-solving mechanisms' as well as tools for gatekeeping and selection, tasked with, in their most often discussed form, 'automated assignment of relevance to certain selected pieces of information' (Just and Latzer 2017: 239; see also Alpaydin 2016: 16) and making them 'meaningfully visible' (Bucher 2018: 3). In societies characterised by overabundance of choice in essentially all domains of life, from shopping to dating to voting, choice itself becomes a challenge, threatening to 'overwhelm individuals and undermine the benefits that choice can provide' (Graham 2018: 1–2). Hence, automated choice (or at least recommendation), sorting out of the environment to be experienced, and provision of an architecture within which action is to take place can seem a salvation to many.

Of course, it must always be remembered that in a way, 'algorithms are simplifications, or distortions' that 'take a complex system from the world and abstract it into processes that capture some of that system's logic and discard others' (Bogost 2015). However, instead of being a limitation of the power of algorithms, this selection, simplification, and amplification is one of the *sources* of their power. Instead of working on the world as it is, algorithms strengthen a particular version of it, serving the interests of the algorithm-wielding actors and narrowing down the set of potentially problematic issues for the consumers (understood in the broadest sense possible) by serving them a version of reality already prescribed by those in control of the code. In popular use, though, the term has taken on the role of a generic shorthand for any calculative computer system with the capacity to make decisions more quickly, accurately, and comprehensively than any human being would be able to (Beer 2019: 11). The preceding characteristics clearly pinpoint several major sources of algorithmic power and also give currency to Kemper and Kolkman's (2018: 2) assertion that we now live an 'algorithmic life', i.e. life that is structured and shaped by and through algorithms. And while it is not completely unfounded for Bogost (2015) to claim that '[c]oncepts like "algorithm" have become sloppy shorthands for mistaking multipart complex systems for simple, singular ones', it is nevertheless also true that, properly understood, algorithms do merit a role at the very epicentre of today's life.

The sheer availability of data upon which algorithmic decisions can be made, and the potential for generating even more, means that there are, by now, algorithms lurking behind almost every conceivable aspect of life, such as algorithms governing the recommendation of songs on a streaming service, market prices of commodities, the seating of customers at a restaurant, the matching of potential partners on dating apps, the serving of personalised apps, etc.—the list could be next to endless (Greenfield 2018: 212). Simultaneously, it is crucial to note that algorithms, just like other digital technologies, are inextricably *social*, i.e. 'they are both socially *produced* and socially *productive* of particular effects', including, among other things, 'new ways of doing things, new forms of social and economic relations, new modes of cultural activity, and new ways of exchanging information and producing knowledge' (Williamson 2017: 267). Although it would not be entirely incorrect to imagine algorithms supporting and upholding the shape of our social world similarly to, in physical architecture, a frame upholding the structure and shape of a building, algorithms still constitute very peculiar frames, nevertheless. These are shapeshifting protean frames that simultaneously react and adapt while also conditioning the shape and form of others.

Clearly, while on the one hand, algorithms are clearly concerned with 'the orchestration of existence' (Langlois and Elmer 2018: 10), that orchestration is based on an original score composed out of previous interactions which include interactions within the world as well as previous world-algorithm interactions. Or, to put it another way, algorithms are to be conceived of in a way similar to a Möbius strip—as something that has neither pure inside nor pure outside. The preceding is a necessary clarification when discussing the role of algorithms in order to avoid falling for what Bogost (2015) calls 'the worship of the algorithm': assigning God-like power to code while ignoring its dependence on broader assemblages (the latter are delved into more deeply in the chapters that follow). Still this complication of power does not deny (and this is largely overlooked by critics such as Bogost) the fact that, debates about the precise degree of permeation notwithstanding, the specificity of today's society and power relations is precisely that they are permeated by algorithms, thereby allowing us to talk of an algorithmic culture and society nevertheless.

Through application of machine learning techniques, it has become possible to identify 'trends, relationships, and hidden patterns in disparate groups of data' as algorithms 'adapt their code and shape performance based on experience' (Perel and Elkin-Koren 2017: 189). This quality is particularly important in situations when a single algorithm for solving a problem is not yet known or is, perhaps, unfeasible to write, thereby implying the need for a learning algorithm (Alpaydin 2016: 16–17). In other words, algorithms become interactive, establishing a two-way relationship with the broader context in which they operate, learning from the world and then feeding back to the world by means of sorting, ranking, and selection, then learning again from the world that has been affected by their own previous output, then further feeding back, and so on. As a result, algorithms that possess a learning capacity of their own go beyond being mere tools used for the attainment of their designers' and implementers' goals—'they effectively shape the meaning of the goals themselves' (Perel and Elkin-Koren 2017: 189).

In this sense, the question of agency, further elaborated in Chapter 5, becomes crucial: if even the implementers become dependent upon the implemented, does that flip agency, at least in part, from humans to code?

A further aspect of the social nature of algorithms lies in their significant impact upon the social and cultural environment, such as shaping the content and the flow of information that users encounter online, particularly on social media, the inclusion, distribution, and ranking of search results, or recommendation services on online retail websites, thereby, once again, sifting through and learning from the available data while simultaneously strongly affecting the conditions for further emergence and generation of such data, e.g. by affecting user opinions, decisions, actions, etc. (Dourish 2016: 7; see also Burrell 2016). Hence, algorithms arrogate to themselves a significant portion of agency in affecting the substance and nature of our life-worlds, not only in terms of incremental everyday conditioning but also with regards to fundamental shifts, simply because 'changes to any of the more widely relied-upon algorithms can have consequences that ripple through the entire society', particularly because 'certain perspectives on reality are reinforced, and others undermined' (Greenfield 2018: 212). In this sense, code can be seen as even more deterministic than law: If in legal systems (at least democratic ones) actions that are not prohibited are typically considered to be allowed, in systems structured through code only actions that are allowed are possible. In effect, this leads to 'programmed forms of sociality' in which the shape and form that human interactions and relationships assume is pre-written in code (Bucher 2018: 2), acting as an opaque form of regulation. One can, therefore, observe a clear mismatch between the ordering role of the algorithms and their opacity which often renders them imperceptible to the general public (Cotter 2018: 2).

Moreover, algorithms are social because they are inseparably intertwined with our value systems by being both influenced by human value structures at their design and learning stages and generative of such values. To begin with, algorithms are not impartial, objective, and value-neutral—on the contrary, they are 'inescapably value-laden' because their '[o]perational parameters are specified by developers and configured by users with desired outcomes in mind that privilege some values and interests over others', at the very least in terms of choosing one of the many possible design options, variables and their values, and so on (Mittelstadt et al. 2016: 1; see also Bucher 2018). As stressed by

Greenfield (2018: 233), '[t]he choices we make in designing an algorithm have profound consequences for the things that are sorted by it' to the extent that '[e]ven the choice of weighing applied to a single variable can lead to different effects in the application of an algorithmic tool'. As a result, the decisions that algorithms make are themselves inescapably biased, their functionality and design reflecting 'the values of its designer and intended uses' (Mittelstadt et al. 2016: 7). In that sense, algorithms can be paradoxically human-like.

In fact, algorithms can be seen as non-objective in two major ways: One is the aforementioned internal bias (as inscribed in code) while the second is an external one, moulded by the environment in which they end up operating. As Kelleher and Tierney (2018: 191) pinpoint, when operating in an environment characterised by distinct biases, data-crunching algorithms end up basing their outputs on the said biases, and the more socially pronounced the biases, the stronger and more consistent the bias-derived outputs become (see also Janssen and Kuk 2016: 372). In a similar fashion to Horkheimer's (2013) critique of the instrumentalization of reason that only serves to disguise ideology, emphasis on supposed algorithmic neutrality in performing processes and making predictions and decisions could, instead, serve to mask the interests, biases, and ideological propensities of the code-writers (or their patrons), disguising them as somehow natural and objective. Indeed, Bogost (2015) is correct in his claim that by being focused on algorithms as independent rational actors we run the risk of turning computers (although code would be more appropriate here) into gods and their outputs into scripture. As a result, it would be wrong to look at algorithmic or datafied governance as a means of replacing the biases and subjectivities of human governing agents with something rational and objective—that would only mean a replacement of one partial ordering of the world with another, just better cloaked this time.

As clear from the above, it is right to assert, as Sunstein (2018: 3) does, that '[w]e live in the age of an algorithm, and the algorithm knows a lot': the changes are profound enough to necessitate a break in thinking and periodisation, and the knowledge about each and every one of us that can be teased out by an algorithm exceeds our own knowledge about ourselves, our friends and family, and the wider world. And if such knowledge is still lacking, it can be reasonably inferred from the other bits and pieces of information that are already available: '[i]f the algorithm knows that you like certain kinds of music, it might know, with a high probability, what kinds of movies and books you like, and what political candidates will appeal to you' and, by conflating that with the knowledge of, e.g. websites visited, 'it might well know what products you're likely to buy, and what you think about climate change and immigration', ultimately making fine-grained distinctions even between otherwise outwardly similar individuals (Sunstein 2018: 3–4).

Moreover, the accumulation and articulation of granular knowledge about individuals is further enabled by the fact that most activities through which data are collected are of the algorithms' own making since '[a]lgorithms increasingly define the spaces of our information encounters, encounters with others, and the status of knowledge as it is produced and circulates in digitally-mediated contexts' (Boler and Davis 2018: 82–83). In that sense, it is not difficult to comprehend why Andersen (2018) postulates our world to be structured around the paramount building blocks of databases that organise and store data and the algorithms that carry out the functions of filling such databases, ranking and otherwise making sense of their content, and making such content searchable. Such is the architecture of ordering in today's world.

Matters get further complicated, though, in the context of machine learning and algorithms that are dynamic, capable of adapting to the environment and, at least to an extent, reprogramming themselves, thereby eliding the traditional onus on the designer to take responsibility for the choices made (and not made) while writing the code; instead, a code that is fluid and largely context-dependent precludes the intelligibility and oversight of its operating practices, decision-making pathways, and even reasons for the presence (or absence) of the pathways in question (Mittelstadt et al. 2016: 11). As a result, despite the impression that automation works in a neutral fashion, it is by no means a deviation but, instead, a rule that algorithms and their selection mechanisms can potentially operate in ways that are in some respect discriminatory (Andrews 2018: 4; Duguay 2018: 29). That fundamentally needs to be taken into account when discussing the regulatory power of algorithms.

The abovementioned power becomes even more acute when one considers the use of algorithms not only for categorising and sorting consumers and other target audiences but also for modulating attention allocation strategies and shaping encounters and interactions (Carah 2017: 396). In fact, publics (as bearers of will and subjects of governance) themselves only come about 'when technologies create associations by aggregating people' (Annany 2016: 100). In other words, algorithmic sorting and filtering creates groups where previously were none (or different groups existed). It is entirely plausible that none of these newly created publics would have ever come to existence in the first place if it was not for algorithms and their capacity to detect salient traits pertinent to the end user of data as well as establishing nonobvious connections that nevertheless relate people, perhaps even without them noticing the connection having taken place.

On the other hand, connections can be manufactured not through detection of pre-existing traits but by inciting action or, at the very least, thought, i.e. through the capacity of algorithmic agenda setting (thus determining *what* we end up thinking about) and framing (*how* we approach those agenda items), which collectively serve to determine how we act (Just and Latzer 2017: 245). As '[a]ttention is drawn to certain things at the expense of others', individuals' consciousnesses and perceived realities are affected in profound ways (Just and Latzer 2017: 245–246); and when these ways coalesce, new publics are formed. In this context, it is not surprising that Bucher (2018: 3) comes to describe algorithmic power in primarily Foucauldian terms: as productive of particular, pre-programmed 'forms of acting and knowing' that manifest themselves through the effects of the code.

In addition (or, perhaps, as a corollary) to structuring the information and choice environment as well as the very experience of environment encountered by individuals, algorithms also act as complex profiling tools, neatly sorting individuals into categories, groups, and niches for purposes ranging from placement of advertisements to political marketing to law enforcement. Such profiling also helps individuals maximise their own impact on the environment since the attributes algorithmically assigned to a person's profile can allow intelligent systems to make choices, carry out operations, and make sense of the results, subsequently feeding back to the ever-increasing personalisation of one's lived environment (Mittelstadt et al. 2016: 3). Such ever-increasing personalisation is indeed key: it is part and parcel of algorithmic governance that it only successfully operates by offering us what we want even on occasions when we do not yet consciously know that we wanted it (or did not know prior to it being offered); hence, what we ultimately become engaged with is likely to be not what is important but what we are going to like (Carlsson 2017: 11). Such an environment is definitely more pleasurable and enjoyable than the one in which classic causality operates, and that is why the algorithmically ordered condition seems so attractive. Nevertheless, by subscribing to this condition (provided that there is a possibility *not to* subscribe or to unsubscribe), we ascribe ourselves to a paradoxical condition: The world is simultaneously structured around us (based on algorithmic determination of our likes and desires) and impenetrable to us, responsive not to us but to a power beyond us.

As a further matter, algorithmic discretion progressively exceeds the mundane, as manifested by mediation and dispute resolution algorithms, policing algorithms that predict potential hotspots for criminal activity and, perhaps, potential suspects prior to them even committing a crime, or algorithms that aid clinical decision-making up to the level of suggesting the likely diagnoses and potential treatments (Mittelstadt et al. 2016: 3). In the latter case at least, algorithmic choices can literally be a matter of life and death. In other words, algorithms can easily be seen as mechanisms of governance, or power-exerting tools, wielded by groups of individual actors and institutions, the state included, but increasingly acquiring autonomous power of their own (Just and Latzer 2017: 245). It can, therefore, be stated that, in the context of ever further digitisation, algorithmic processes are 'moving to the heart of the governance of our society' because '[i]mportant decisions about people are increasingly made by algorithm' (Janssen and Kuk 2016: 371). However, the shift is broader than merely one involving the means of governance. The outlook of the latter has changed as well, turning into a notably technocratic one, which 'assumes that complex societal problems can be deconstructed into neatly defined, structured and well-scoped problems that can be solved algorithmically and in which political realities play no role', thereby aspiring to neutrality, objectivity, and therefore, legitimacy (Janssen and Kuk 2016: 371–372). Hence, it is the task of the section below to critically analyse this new regime of governance.

3.2   Governance and Law

As asserted by Danaher et al. (2017: 1–2), '[w]e are living in an algorithmic age where mathematics and computer science are coming together in powerful ways to influence, shape and guide our behaviour and the governance of societies' through the use of algorithms to 'nudge, bias, guide, provoke, control, manipulate and constrain human behaviour'. In the broadest sense, algorithmic governance can be defined as 'the automated collection, aggregation, and analysis of big data, using algorithms to model, anticipate, and pre-emptively affect and govern possible behaviours' (Williamson 2017: 271) or, similarly, 'the ways in which our digitally-mediatized experiences of the world [...] are shaped by artificial intelligence of algorithms designed according to commodified, consumer-oriented logics' (Boler and Davis 2018: 83). However, the 'consumer-oriented' nature of algorithmic governance, set out in the latter definition, needs further elaboration. While not entirely false (such governance is consumer-oriented in the sense of aiming to achieve the greatest consumer satisfaction possible in order to hook them on a service or platform and thereby convince them to surrender as much data as possible), it might give a false impression that consumer orientation is a goal in itself. Instead, as hinted in the preceding, the actual end is the extension of corporate power and influence and, therefore, revenue.

In its enactment, algorithmic governance marks a shift from a norm-based disciplinary logic to a logic of control, i.e. one based on opening and closing different possibilities depending on the interests behind the algorithms; for the latter, it suffices to create a terrain on which individuals, with characteristics that they *already* have, are bound to 'freely' choose a pre-set outcome (Carah 2017: 396). As a result, there is no more need to actively *shape* individuals in disciplinary terms, as was the case, perhaps most notably, for Foucault (2012). Crucially, then, algorithms 'embody values and can organize and impose order on society by both affording and impeding certain practices, behaviors, and activities' (Just and Latzer 2017: 246). In fact, algorithmic selection has become so prominent that, arguably, societies as such are 'increasingly being co-constructed' by it, particularly as people's behaviour is influenced through algorithmic management of perception (Just and Latzer 2017: 254). Under such circumstances, algorithmic conditioning and, thereby, governance is reduced to a matter of assigning the appropriate weightings to variables laid out in code.

Although algorithmic governance, due to its primarily corporate and/or technical character, might not necessarily be intuitively seen as political, it concerns matters at the heart of the political par excellence. In particular, the 'politics' here is understood as concerning 'ways of world-making', i.e. 'the practices and capacities entailed in ordering and arranging different ways of being in the world' or, perhaps even more radically, 'the making of certain realities' (Bucher 2018: 3). In other words, the politics of algorithmic governance is 'politics of the real, of what gets to be in the world' (Bucher 2018: 3). Therefore, the matter becomes one of inclusion/exclusion, but not merely in terms of standpoints and values but also with regards to the constitutive material and social conditions of the latter. There is, however, a crucial caveat to keep in mind here: the corporate character of algorithmic governance. Whereas, in democracies at least, politics assumes the form of public contestation over regulation, distribution, and allocation, algorithmic governance structures the reality of the sociopolitical domain without being contested in that same domain.

Through algorithmic governance, we have politics that is still political in its function and effect but not at its premise. This shift, then, has a crucial effect on the value considerations that go into the making of decisions that are essentially political. Whereas, in an ideal-case scenario at least, democratic politics is about articulation of public interests and demands, algorithmic governance-qua-politics is 'folded into promises of profit and business models', meaning that 'a "good" and well-functioning algorithm is one that creates value, one that makes better and more efficient predictions', making users engage and, therefore, return to the platform as frequently as possible (Bucher 2018: 6). Such goals are, of course, not necessarily in line with any articulations of public demands (instead, they are more likely to be formative of such demands)—a clear contrast to what we normally associate with 'good' forms of governance and good modes of regulation (and, as argued below, law).

A clear problem with algorithmic governance is its general lack of transparency. It is indeed alarming that while 'corporate actors have unprecedented knowledge about the minutiae of our daily lives', we ourselves 'know little to nothing about how they use this knowledge to influence the important decisions that we – and they – make' (Pasquale 2015: 9). This non-transparency is inscribed in the very nature of algorithms as 'their decision-making criteria are concealed behind the veil of code that we cannot easily read and comprehend' (Perel and Elkin-Koren 2017: 181). Moreover, that complexity and opacity could have even deeper antecedents in 'the technical necessity of handling the complexity of the system', including its sheer scale and typically multilayer nature (Bucher 2018: 41, 57). Indeed, in the context of algorithmic governance, or regulation through code, we are permanently faced with 'algorithmic unknowns', particularly in the context of machine learning and the ensuing constant evolution of the relevant algorithm, as well as the structural complexity of the entire edifice, which ultimately makes the key life-structuring algorithms tortuous and impenetrable (Andrews 2018: 6).

Of course, relative impenetrability of key ordering tools could be said to be not new: For example, legal codes, exerting strong regulatory power, may often also be difficult to comprehend, yet alone interpret, without at least some degree of specialist knowledge. However, computer code adds a further layer of impenetrability, and the more complex the code is, the more specialist knowledge is necessary. Moreover, even if the meaning was deciphered at one given point in time, the same knowledge may not necessarily remain accurate in the future because the more efficient of today's algorithms are capable of dynamically evolving as new and different data patterns are discovered or are immediately modified by their creators (Perel and Elkin-Koren 2017: 181; see also Greenfield 2018: 253). But there is even further potential for complexity: one can only imagine how difficult reconstructing the reasons for a particular decision would be in cases when multiple algorithms are involved simultaneously, each contributing an unspecified and continuously varying bit towards the outcome and interacting in multiple, non-linear ways (Greenfield 2018: 253). In fact, such situations are not exceptions but the norm. In most of the advanced and dominant platforms, 'there is no such thing as *the* algorithm but, instead, a multiplicity of task-specific algorithms', the multiplicity of which automatically implies perpetual change, rendering such platforms constant work in progress (Bucher 2018: 47–48).

Even if the preceding problems were to be overcome, for example by constant monitoring, the very access to code is often forbidden as 'algorithms that enforce online activity are often implemented by private, profit-maximizing entities, operating under minimal transparency obligations' (Perel and Elkin-Koren 2017: 181). In fact, not only such obligations are minimal—transparency as such goes against the business models of the companies involved, and algorithms around which their entire businesses are structured end up being tightly guarded commercial secrets. As a consequence, our society can appropriately be called a 'black box society', referring to the clear paradox that although 'data is becoming staggering in its breadth and depth' simultaneously 'the information most important to us is out of reach, available only to insiders' (Pasquale 2015: 191; see also Alpaydin 2016: 154). As argued later in the book, such changes render ascription of agency to human actors problematic.

Alternatively, others would suggest that due to the constantly changing nature of algorithms and their permanent multiplicity and interaction they should be seen as 'simply neither black nor box but eventful'

(Bucher 2018: 48). Nevertheless, the two renderings are not necessarily mutually exclusive: what they ultimately refer to is, in both cases, the impossibility of comprehending the modes and tools of governance one is subjected to. In effect, then, in this world of data-driven decisions and predictions 'we may not be able to explain the reasons behind our decisions' (Mayer-Schönberger and Cukier 2017: 17), in stark contrast with the public nature of conventional decision-making, especially by public authorities. On the other hand, if greater transparency merely shifts the onus on individuals, whose resources in terms of power, time, and knowledge are already limited and permanently stretched ever further, as Annany and Crawford (2018) warn, then mere transparency could make things worse, not better.

A key problem has thus been laid bare: while the offline life is regulated by promulgated laws, the online environment is regulated by opaque code. However, to complicate matters even further, the very line between offline and online is becoming increasingly blurred, with the online progressively encroaching on the offline. Such an encroachment is rendered particularly visible through the rise in prominence of 'cyber-physical infrastructure' that denotes 'turning devices, homes, public and private transport, bridges, hospitals and offices online, to enable persistent monitoring and surreptitious adaptation' (Hildebrandt 2016: 4). Crucially, such monitoring is not necessarily synonymous with malevolent surveillance (although the danger of it becoming such always lurks in the background)—instead, such monitoring should be seen as enabling dynamic change within our environment in line with our preferences, uses, and tastes (through continuous feedback loops in which user behaviour is fed back into algorithmic infrastructure), filtered through corporate and state interests.

Regulation can, of course, take different forms. Regulation by law is perhaps, the most straightforward—it provides for punishment for some things and incentives for others; however, there are other modes of regulation as well: norms determine the social cost and utility of actions in accordance with a community's cultural specificities while markets incentivise activities that are profitable and disincentivise the costly ones while technological architecture literally enables and disables activities (Peters and Johnson 2016: 63–64; see also Lessig 2000: 508–509). Indeed, as Lessig noted a while ago already, while we know very well how 'constitutions, statutes, and other legal codes' regulate the physical environment, in the same vein 'we must understand how a different "code" regulates –

how the software and the hardware (i.e., the "code" of cyberspace) that make cyberspace what it is also regulate cyberspace as it is' (Lessig 2006: 5). Hence, as his famous dictum goes, 'code is law' (Lessig 2006: 5; see also Lessig 1999) because it sets the features of the online environment that 'constrain some behavior [...] by making other behavior possible'; in addition, the coded features 'embed certain values, or they make the realization of certain values impossible' (Lessig 2000: 510). Of course, the malleability of the online environment allows for a broader expanse and creativity of regulation than it is the case in the physical world.

While the scope of the original assertion was merely that code regulates the *online* environment by defining 'the terms upon which cyberspace is offered' (Lessig 2006: 84), the progressive erasure of a border between the online and the offline worlds means that regulation through code spills over into the physical environment as well. Moreover, in addition to the general overlap between the two environments, law also becomes dependent on code not only as a means of the former's enforcement but also for elaboration and informational inputs through data collection and analysis (Weber 2018: 701). In the new regulatory environment that opens up for governments, 'citizens become knowable, traceable and trackable across lifespans, social and professional networks, government interactions and geography in new ways as citizens are transformed into "data subjects"' that can even be acted upon pre-emptively if necessary (Redden 2018: 3).

Under conditions of algorithmic governance, as stressed by Greenfield (2018: 212), 'a very great deal of material power reposes in the party that authors the algorithm'. Crucially, then, 'code writers are increasingly lawmakers' determining the default settings and features of the online (and, though it, offline) environment, modes of protecting privacy and anonymity, granting of access, etc. (Lessig 2006: 79; see also Hildebrandt 2016). Certainly offline regulators do not rescind their role completely (European Union's General Data Protection Regulation being a notable example) but even then the law can only provide a framework while the actual offering is in the hands of those in control of code. After all, the online architecture, constructed through code, 'materially influences human behaviour', and '[b]ecause code can achieve a nearly perfect control in cyberspace, architecture becomes the most powerful regulator' (Weber 2018: 703). Indeed, everyday activities now take place against the backdrop of 'digital platforms directing and limiting action by providing a "grammar of action" that makes certain activities doable' while discarding others (Törnberg and Törnberg 2018: 8).

In addition, algorithms should be seen as exercising a disciplinary function by determining who and what gets seen, under what conditions, and in exchange for what behaviour, which is a clear regulatory role (Cotter 2018: 2). Such regulation extends beyond the conditioning (subtle or not so much) of natural persons but also reaches deeply into the business environment, becoming a fundamentally influential force there as well. Rather straightforwardly, '[e]very time Google tweaks its search algorithm, or Facebook the one it uses to govern story placement, certain business propositions suddenly become viable, and others immediately cease to be' (Greenfield 2018: 212). That is a manifestation of regulatory power more potent and more immediate than much of the traditional law could aspire to. Even more important, though, is the fact that often such an immediate and potent effect is a side-effect: while legal regulation is typically enacted with the explicit aim of ordering a particular sphere of human activity by public institutions under a public and predefined procedure and, therefore, open for scrutiny, algorithmic regulation can be a mere corollary to a private company disposing of its private property (algorithm) as it wishes behind closed doors with the aim of maximising private returns and without even foreseeing or taking into consideration the multiple effects that are going to ripple around in the wake of even a minor tweak.

Indeed, the stability—change axis constitutes a further notable tension between conventional law and algorithm-as-law. Whereas conventional law typically enjoys (relative) stability, and such stability is typically seen as a value (see, for example, Macdonald and Atkins 2010; Ávila 2016; Fenwick and Wrbka 2016; Fenwick et al. 2017), algorithmic law is characterised by a spirit of perpetual experimentation, the latter being 'part and parcel of how most online platforms now develop their products' (Bucher 2018: 48). Such experimentation typically assumes the form of so-called A/B testing that enables the digital architects-qua-lawmakers to compare different versions of their algorithms by simultaneously subjecting different groups of users to different versions of it; while in principle this might be akin to a focus group, the difference is that there is no participant consent, informed or otherwise (Bucher 2018: 48).

Moreover, the stakes are certainly higher in A/B testing of algorithms that ultimately act as laws in comparison to focus groups: whereas in the latter participants are explicitly asked for feedback on a clearly defined matter and, thus, should be capable of drawing a clear line between research and real-life conditions, in the former it is the *real life* of unknowing and unwilling participants that is being experimented upon. As a result, when delving into matters of algorithmic governance, 'the question inevitably arises as to which version, what test group, or what time frame we are talking about' (Bucher 2018: 48). In traditional law, meanwhile, no such presence of multiple simultaneous versions of regulation and, therefore, reality is possible. That is a matter not only of feasibility but also of principle—if such experimentation takes place and multiple co-present versions of regulation are in force, equal treatment of persons is impossible.

Writing in 2006, Lessig could only speculate that 'a code of cyberspace, defining the freedoms and controls of cyberspace, will be built'—the only question was 'by whom, and with what values' (Lessig 2006: 6). Such code is now, perhaps, being finalised, and we, in all likelihood, are no longer able to even make a choice about it—at least not one unaffected by that same code. In this vein, as Vaidhyanathan (2018: 99) correctly observes, the current stage of competition between the technology giants is no longer over being 'the operating systems of our laptops and desktops' but, instead, 'to be the operating system of our lives'. This development is the likely outcome of the network effects discussed above: as platforms expand and their owners branch out into ever more distinct areas of life, acquiring companies operating in industries anywhere from smart appliances to self-driving cars to virtual reality, the data collected, the services offered, and the ability to tailor any offering become so vast that any smaller competitors simply do not stand a chance. The only question is which of the current technology giants is going to successfully monopolise the market, becoming such an operating system in itself.

While for Lessig the greatest threat seemed to be state regulation, perhaps with code-writers' complicity due to a shared interest in a more controllable online environment (see Lessig 2006: 72–80), it is, in fact, the private business interest enshrined in code that has become the paramount shaper of today's choice architecture, with the potential of subsuming all activities under itself. Hence, while 'a world in which regulation by code is not easy' (Zittrain 2006) might be preferable, it simultaneously appears to become ever less realistic. In this context, it seems naïve of Lessig (2006: 200) to have proposed more choice of private code as a solution: while it might appear that if choice exists and products compete on how to better reflect consumer preferences then the power of corporations is restricted, as shown above, reflecting user preferences is best achieved through monopolisation of the market and, therefore, user data, thereby exerting a strong push towards concentration (see also Mayer-Schönberger 2008: 721).

The above shift towards a monopoly of code and data, once it happens, will imply the construction of not only *the* definitive code of cyberspace that acts as its fundamental constitutional law but simultaneously also *the* code that structures everyday life as well in all its experiences and affordances, both physical and virtual—a constitutional code that transcends boundaries and divides. This *constitutional code of all spaces* would not only set the *Grundnorme* in the Kelsenian sense (see Kelsen 1989, 1999) through its algorithmic 'if... then' rules but also enact and interpret them as well as further constitutional laws as well as laws and regulations of lesser standing, thus moving from mere commercial enterprise to a self-executing system of code-as-law. But this 'operating system of our lives' is likely to go even further: In addition to setting the limits of the actionable, the sayable, the experienced, etc., it will have access to (and, therefore, will collect and analyse) the entirety of our data relating to any of our 'activities and states of being', thereby becoming capable of imperceptibly guiding our decisions and actions (Vaidhyanathan 2018: 99). In other words, code will be more than just law: it will not only set binding rules but will also nudge us in prescribed directions so that we only take some of the pathways made available through the architecture of rules—and these pathways will be personalised if necessary.

Here two different economic renderings of law should be taken into account as an illustrative example. Under the neoclassical model of economic analysis of law (see, notably, Posner 2014), laws are presumed to reflect efficient solutions to organisational problems. While this model strongly promotes the standard economic idea of 'consumer sovereignty', assuming that individuals always know best and make rational choices, its alternative, a behavioural economics-inspired view of law calls that key premise into question by focusing on bounded human rationality instead: for them, both ordinary citizens and government officials (as sources of law and regulation in the broad sense) are fallible and susceptible to motivational and cognitive problems (Jolls et al. 1998: 1541). As a result, benevolent nudging of a libertarian paternalist nature (encouraging individuals to make choices preferred by the regulator while not eliminating alternatives) starts making sense. Nevertheless, a difference in case of regulation by code-as-law must be stressed again as here such benevolent paternalism is unlikely. Instead, it is more than likely that bounded rationality will be abused, particularly because such regulation will be not by public actors seeking to create public goods but by private actors seeking to create private goods for themselves. And even in cases of non-bounded rationality, the malleability of the online environment and its openness to rebuilding through the very code that regulates and serves as law, implies that the choices available to even the rational sovereign consumer can be designed in a way that leads to an otherwise troublesome conclusion that favours the regulator and not the regulated. Thus, both the behavioural and the standard models face serious shortcomings in the digital environment.

References

Alpaydin, E. (2016). *Machine Learning: The New AI*. Cambridge, MA and London: The MIT Press.

Andersen, J. (2018). Archiving, Ordering, and Searching: Search Engines, Algorithms, Databases, and Deep Mediatization. *Media, Culture and Society*. https://doi.org/10.1177/0163443718754652.

Andrews, L. (2018). Public Administration, Public Leadership and the Construction of Public Value in the Age of the Algorithm and 'Big Data'. *Public Administration*, Published Online Before Print on 6 August 2018, https://doi.org/10.1111/padm.12534.

Annany, M. (2016). Toward an Ethics of Algorithms: Convening, Observation, Probability, and Timeliness. *Science, Technology and Human Values, 41*(1), 93–117.

Annany, M., & Crawford, K. (2018). Seeing Without Knowing: Limitations of the Transparency Ideal and Its Application to Algorithmic Accountability. *New Media & Society, 20*(3), 973–989.

Ávila, H. (2016). *Certainty in Law*. New York: Springer.

Beer, D. (2019). *The Data Gaze: Capitalism, Power and Perception*. Los Angeles and London: Sage.

Bogost, I. (2015, January 15). The Cathedral of Computation. *The Atlantic*. Available at https://www.theatlantic.com/technology/archive/2015/01/the-cathedral-of-computation/384300.

Boler, M., & Davis, E. (2018). The Affective Politics of the 'Post-Truth' Era: Feeling Rules and Networked Subjectivity. *Emotion, Space & Society, 27*, 75–85.

Bucher, T. (2018). *If... Then: Algorithmic Power and Politics*. Oxford and New York: Oxford University Press.

Burrell, J. (2016). How the Machine 'Thinks': Understanding Opacity in Machine Learning Algorithms. *Big Data and Society.* https://doi.org/10.1177/2053951715622512.

Carah, N. (2017). Algorithmic Brands: A Decade of Brand Experiments with Mobile Social Media. *New Media & Society, 19*(3), 384–400.

Carlsson, M. (2017). Automating Judgment? Algorithmic Judgment, News Knowledge, and Journalistic Professionalism. *New Media and Society, 20*(5), 1755–1772.

Cotter, K. (2018). Playing the Visibility Game: How Digital Influencers and Algorithms Negotiate Influence on Instagram. *New Media & Society.* https://doi.org/10.1177/1461444818815684.

Danaher, J., et al. (2017). Algorithmic Governance: Developing a Research Agenda through the Power of Collective Intelligence. *Big Data & Society.* https://doi.org/10.1177/2053951717726554.

Dourish, P. (2016). Algorithms and Their Others: Algorithmic Culture in Context. *Big Data & Society.* https://doi.org/10.1177/2053951716665128.

Duguay, S. (2018). Social Media's Breaking News: The Logic of Automation in Facebook Trending Topics and Twitter Moments. *Media International Australia, 166*(1), 20–33.

Fenwick, M., Siems, M., & Wrbka, S. (Eds.). (2017). *The Shifting Meaning of Legal Certainty in Comparative and Transnational Law.* Oxford and Portland: Hart Publishing.

Fenwick, M., & Wrbka, S. (Eds.). (2016). *Legal Certainty in a Contemporary Context: Private and Criminal Law Perspectives.* New York: Springer.

Foucault, M. (2012). *Discipline and Punish: The Birth of the Prison.* New York: Vintage Books.

Graham, T. (2018). Platforms and Hyper-Choice on the World Wide Web. *Big Data & Society.* https://doi.org/10.1177/2053951718765878.

Greenfield, A. (2018). *Radical Technologies: Thee Design of Everyday Life.* London and New York: Verso.

Hildebrandt, M. (2016). Law *as* Information in the Era of Data-Driven Agency. *The Modern Law Review, 79*(1), 1–30.

Horkheimer, M. (2013). *Eclipse of Reason.* London and New York: Bloomsbury Academic.

Janssen, M., & Kuk, G. (2016). The Challenges and Limits of Big Data Algorithms in Technocratic Governance. *Government Information Quarterly, 33,* 371–377.

Jolls, C., Sunstein, C. R., & Thaler, R. (1998). A Behavioral Approach to Law and Economics. *Stanford Law Review, 50,* 1471–1550.

Just, N., & Latzer, M. (2017). Governance by Algorithms: Reality Construction by Algorithmic Selection in the Internet. *Media, Culture and Society, 39*(2), 238–258.

Kelleher, J. D., & Tierney, B. (2018). *Data Science*. Cambridge, MA and London: The MIT Press.

Kelsen, H. (1989). *Pure Theory of Law*. Gloucester, MA: Peter Smith.

Kelsen, H. (1999). *General Theory of Law and State*. Union, NJ: The Lawbook Exchange.

Kemper, J., & Kolkman, D. (2018). Transparent to Whom? No Algorithmic Accountability Without a Critical Audience. *Information, Communication & Society*. https://doi.org/10.1080/1369118x.2018.1477967.

Langlois, G., & Elmer, G. (2018). Impersonal Subjectivation from Platforms to Infrastructures. *Media, Culture and Society*. https://doi.org/10.1177/0163443718818374.

Lee, M. K. (2018). Understanding Perception of Algorithmic Decisions: Fairness, Trust, and Emotion in Response to Algorithmic Management. *Big Data & Society*. https://doi.org/10.1177/2053951718756684.

Lessig, L. (1999). *Code and Other Laws of Cyberspace*. New York: Basic Books.

Lessig, L. (2000). The Law of the Horse: What Cyberlaw Might Teach. *Harvard Law Review, 113*, 501–549.

Lessig, L. (2006). *Code: Version 2.0*. New York: Basic Books.

Macdonald, E., & Atkins, R. (2010). *Koffman & Macdonald's Law of Contract* (7th ed.). Oxford and New York: Oxford University Press.

Mayer-Schönberger, V. (2008). Demystifying Lessig. *Wisconsin Law Review, 4*, 713–746.

Mayer-Schönberger, V., & Cukier, K. (2017). *Big Data: Th Essential Guide to Work, Life and Learning in the Age of Insight*. London: John Murray.

Mittelstadt, B. D., et al. (2016). The Ethics of Algorithms: Mapping the Debate. *Big Data & Society*. https://doi.org/10.1177/2053951716679679.

Pasquale, F. (2015). *The Black Box Society: The Secret Algorithms That Control Money and Information*. Cambridge, MA and London: Harvard University Press.

Perel, M., & Elkin-Koren, N. (2017). Black Box Tinkering: Beyond Disclosure in Algorithmic Enforcement. *Florida Law Review, 69*(1), 181–221.

Peters, J., & Johnson, B. (2016). Conceptualizing Private Governance in a Networked Society. *North Carolina Journal of Law and Technology, 18*(1), 15–68.

Posner, R. A. (2014). *Economic Analysis of Law* (9th ed.). New York: Wolters Kluwer Law & Business.

Redden, J. (2018). Democratic Governance in an Age of Datafication: Lessons from Mapping Government Discourses and Practices. *Big Data & Society*. https://doi.org/10.1177/2053951718809145.

Sunstein, C. R. (2018). *#Republic*. Princeton and Oxford: Princeton University Press.

Törnberg, P., & Törnberg, A. (2018). The Limits of Computation: A Philosophical Critique of Contemporary Big Data Research. *Big Data & Society.* https://doi.org/10.1177/2053951718811843.

Vaidhyanathan, S. (2018). *Anti-Social Media: How Facebook Disconnects Us and Undermines Democracy.* Oxford and New York: Oxford University Press.

Weber, R. H. (2018). 'Rose is a Rose is a Rose is a Rose'—What About Code and Law? *Computer Law and Security Review, 34,* 701–706.

Williamson, B. (2017). Moulding Student Emotions Through Computational Psychology: Affective Learning Technologies and Algorithmic Governance. *Education Media International, 54*(4), 267–288.

Zittrain, J. (2006). A History of Online Gatekeeping. *Harvard Journal of Law and Technology, 19*(2), 253–298.

CHAPTER 4

Personalisation, Emotion, and Nudging

Abstract Pleasure is undeservedly excluded from studies of algorithmic governance. Nevertheless, due to the incessant competition over attention, prevalent in today's media environment, the maximisation of pleasure and consumer satisfaction becomes a must in order to be able to exert the power of algorithmic governance in the first place. Therefore, the first part of this chapter is dedicated to a discussion of the importance of enthralling one's audience and the role of data therein. The second part, meanwhile, is focused on nudging strategies that are geared towards encouraging individuals to make predefined choices. However, in today's datafied, pleasurised, and personalised environment, nudging goes beyond mere encouragement: As showed in this chapter, options can be stacked in such a way that individuals simply cannot fail to choose the option intended by the choice architect.

Keywords Personalisation · Pleasure · Nudge · Choice · Attention · Architecture

Algorithmic governance takes on many forms. Some of them relate to direct closing and opening of spaces and opportunities, creation and elimination of choice options. Indeed, ranking, sorting, filtering, and creation of digital architecture (i.e. of the very structure and topology of the virtual environment) are, of course, crucial. However, there is also a softer—and decidedly more pleasurable—aspect of algorithmic governance that involves

© The Author(s) 2019
I. Kalpokas, *Algorithmic Governance*,
https://doi.org/10.1007/978-3-030-31922-9_4

nudging individuals in directions preferred by those in control of code. Indeed, the role of pleasure is often overlooked in studies of algorithmic governance and is primarily reserved to literature on the attention economy and affect. Nevertheless, pleasure is to be also seen as part and parcel of successful regulation through algorithms. After all, at least until we have a single operating system of the entire life, algorithmic governance cannot be conceived of as something singular and monolithic. Instead, multiple platforms and other sites of governance have to compete for the attention of the governed-to-be in order to then exercise their power and influence. Once that is achieved, the agents of algorithmic governance can employ nudge strategies in order to encourage the choice options that are deemed to be preferable in advance. And because of the data held, such actors are typically capable of determining which triggers are to be pulled in advance so that the subject of governance cannot fail to choose the predetermined option. However, because of the private nature of algorithmic governance, the question of whose interest is served by such nudge strategies is particularly problematic.

4.1   Pleasure and Satisfaction as an Algorithmic Governance Strategy

As already noted in the previous chapter, the datafication of everyday life and algorithmic analysis of data allow for a great degree of personalisation of ranking, sorting, and digital architecture upon which algorithmic governance is premised, leading subsequently to noticeable 'individualization effects' (Just and Latzer 2017: 254). In other words, not only the content offered to individuals is increasingly personalised but also there is a growing *expectation of* personalised content, and the more personalisation individuals get, the more they expect, in what effectively becomes a personalisation spiral. In practical terms, such a move also implies that datafied offering of personalised content itself becomes generative of the need for ever further accumulation and harvesting of data which is the only means of satisfying the audiences' drive for ever further seamless tailoring.

Such personalisation takes place not only in terms of content but also with regards to emotional and affective flows. This extension should not come as a surprise: after all, '[e]motions and sentiments play a crucial role in our daily lives', assisting in situations such as communication, learning, or decision-making (Poria et al. 2017: 98). That is even more strongly the case in the 'attention economy' of the online environment, in which 'real-time instant gratification' and the ensuing emphasis on consumer satisfaction are key (Evans 2016: 577). After all, the distribution of online content is fundamentally dependent on networks of similarly minded and similarly affected individuals, serving as 'catalysts rather than as professional gatekeepers', i.e. propelling affective content into popularity (Klinger and Swensson 2018: 4). In this way, affective becomes effective.

Naturally, the data that are employed for algorithmic sorting, ranking, and digital environment design involve not only 'facts' but also subjective characteristics, such as 'opinions, sentiments, appraisals, attitudes, and emotions' (Serrano-Guerrero et al. 2015: 18). That is where sentiment analysis and similar techniques come to their own, detecting, selecting, gauging, and interpreting subjective components present in any user-generated content (Etter et al. 2018: 72), which usually comes in the form of an unstructured discourse, rather than a coherent preset narrative (Balazs and Velásquez 2016: 96). The prize is in being able to determine both the actual emotions caused by previous algorithmic choices and the triggers that will need to be pulled in order to elicit (actual or anticipated) satisfaction with the desired outcome of algorithmic governance. Hence, it becomes increasingly apparent that algorithmic governance is concerned not only with architectural affordances but also with the precognitive layer of behaviour. That is in stark contrast to traditional forms of governance that are primarily concerned with constructing the actionable and the unactionable instead of tampering with the emotional preconditions of action (which is something traditional law is largely incapable of doing anyway).

Due to its capacity to successfully determine the emotional state of target audiences, the analysis of opinions and emotions has become a crucial component within diverse fields of activity, including but by no means limited to campaign planning, detection of changes in the ideological setup and patterns within an electorate, or forecasting of financial market performance (Giatsoglou et al. 2017: 214). For this end, machine learning particularly comes to use: As subjective and emotive expressions typically resist being put into predefined expressions and categories, it is only natural to turn to algorithms that possess the capacity of extracting information from and interpreting unforeseen and unexpected data as it emerges in real time (Etter et al. 2018: 74), thereby enabling (almost) automatic maximisation of consumer satisfaction (Puschmann and Powell 2018: 1). Crucially, then, computing is truly becoming affective in terms of the capacity to 'recognize, express, communicate, and respond to humans using emotions' with the ultimate aim of 'generating an affective change in the user' (Schwark 2015: 761, 764). In other words, while conventional regulation primarily deals with classifying and evaluating the actions and omissions of individuals (external power), algorithmic governance by code is more pervasive, capable of *driving* an individual towards acting or not acting (internal power), and thus evoking questions of agency. Since data harvesting can reveal affective triggers that are particularly sensitive and, consequently, the individual *cannot fail to act* on the stimulation, the authorship of subsequent actions becomes unclear, putting the very idea of an autonomous subject in jeopardy.

Clearly, then, the aim is to enable the relevant actors to track largely subliminal flows of affects and emotions and offer products (in the broadest sense possible—from physical artefacts to services to information to experience) and solutions that cater to such flows, thereby increasing product sales and consumer satisfaction (Brigham 2017: 399). And from the consumer perspective, since information is overabundant and thus characterised by 'high velocity and dizzying excess' (Dahlgren 2018: 26), its management and filtering, including that through emotional triggers and nudges, becomes valuable in itself, acting as a deliberately induced choice heuristic (Léveillé Gauvin 2018: 293–294; Vaidhyanathan 2018: 80). In this context, subliminal affective flows become key ways of navigating the environment of information overabundance (Klinger and Swensson 2018: 5). And deliberate tampering with emotional reactions and the underlying affective flows on behalf of the data-rich providers, of course, minimises (or at least significantly reduces) the effort necessary to select the relevant information, make value judgements about it, and rank it by priority. Instead, emotional value is assigned almost automatically and, once registered by a data-mining algorithm, included into one's digital representation for either reinforcement (if the reaction is in accordance with the prescribed aims) or discouragement (if it is not). Once again, the aim of such regulation is to create an intuitive choice pattern that directs audience attention in predictable and programmable ways which the individual (or group) is either known or made to associate with pleasure.

Indeed, algorithmic governance can, and should, be pleasurable. After all, pleasure is not merely a sensation or an emotional state—it is also a bodily condition, leading from a fleeting emotional to a more durable bodily state and increasing both mental *and* physical well-being overall (Damasio 2018: 108–109). As a corollary, then, it must be noted that this inseparability of the body and the mind also implies a physical need for pleasure and satisfaction. A related corollary is also that any stimuli that succeed in hitting the mark have an extra support structure in the body: It is not only associations in one's mind, the electric signals shot between synapses in the brain but also the arousal of the whole body that counts. As Damasio (2018: 159) renders it in a summary of neuroscientific and physiological research, not only subjectivity is a narrative that must be relentlessly constructed but also it originates 'from the circumstances of organisms with certain brain specifications as they interact with the world around, and the world of their past memories, and the world of their interior'. In other words, what we encounter matters because of the way it fits our previous experience and makes the body predispose itself and, therefore, feel in a certain way. Hence, if both the mental and the physical faculties act together, the urge caused by a stimulus must be at least twice as strong, thus further enhancing the power of algorithms capable of identifying precisely the stimulus necessary to induce the required urge and, thus, choice and/or action.

The above potential is further reinforced through possibilities for microtargeting, i.e. direct and strategic transmission of stimuli that are known in advance to sway the recipients in a predefined direction because the stimuli themselves are built upon 'the preferences and characteristics of the individual' (Papakyriakopoulos et al. 2018: 2). As might be expected, microtargeting is a data-intensive process, necessitating the ability to depict not only individual preferences with regards to the particular choice (e.g. political sympathies if it is an election decision that needs to be influenced or desired product associations if a purchasing decision is at stake) but also a much broader spectrum of characteristics that may, directly or indirectly, influence the decision in question (Papakyriakopoulos et al. 2018: 2). Again, if a correct stimulus is directed at the correct target at the correct time, one cannot fail to choose the predetermined option.

While microtargeting, as a practice taken in isolation, perhaps would not imply manipulation per se, such potential is present nevertheless as more and more data are being surrendered and algorithms become ever more precise in their generation and targeting of stimuli, opening up the ability to 'trigger the person's mind to develop a conditioned response necessitated by the algorithm-wielding actor', leading to 'instant influence' through intuitive assimilation of information and action choice (Papakyriakopoulos et al. 2018: 10–11). Such microtargeting capacities take algorithmic governance's focus on emotional proclivities up another level, essentially enabling personalised governance. In contrast to the same laws and regulations applying to everyone equally (as in the conventional rule of law scenario), algorithmic governance enables fine-tuning of triggers and nudges to individual characteristics. As a result, sociality becomes essentially programmed, with humans and non-humans being gathered and associated in predefined ways in accordance with known-in-advance triggers, in all likelihood for the benefit of the programmer of that sociality (Bucher 2018: 4).

Greater capabilities to get to know and interact with individuals, opened up by the digital environment and its largely datafied nature, put an ever-greater emphasis on consumer experience, even giving rise to postulations of a new, experience-based age and an experience-based economy (Riccio 2017; see also Kalpokas 2019). In this new environment, speed and personalisation are key (Colvin and Kingston 2017), implying that a consumer progressively expects to obtain the object of their desire or expectation intuitively, often before they themselves consciously know what that object is. Any query should, therefore, be conceived of as a 'Mirror mirror on the wall' moment when, despite the information sought being seemingly factual, the 'me' is the real issue anyway. In other words, things sought are expected to only further reflect and affirm the enquiring self. Hence, constant and ever-deeper personalisation of consumer experience with the help of algorithms is not merely sensible—it is a must (Colvin and Kingston 2017). Governance is no longer obtrusive or at least no longer perceived as external. It is, instead, turning into experience, wrapping the individual in a dense regulatory layer tightly but gently and pleasurably. That is, essentially, the function of the 'algorithmic' part of algorithmic governance: 'to give us more of what we seemingly want', always being tweaked to make the offering ever more timely, relevant, and exciting (Bucher 2018: 149).

Moreover, attention is a social—communal—thing, structured by 'collective enthralments, which are inextricably architectural and magnetic'; in effect, attention is directed in a particular direction because others are directing attention in that direction—attention is its own magnet

(Citton 2017: 31). As a result, the price of success or failure is high: if one manages to attract the attention of some, that can lead to attraction en masse—partly due to peer pressure (see e.g. Thaler and Sunstein 2009: 59), partly due to the network effects described earlier, and partly because of the use of algorithms to direct attention where others have placed *their* attention, such as recommendations based on what similar individuals have viewed or purchased in the past (Citton 2017: 71). But, since attention attraction is a zero-sum game, any success must imply a corresponding loss for other competing attractors. Consequently, possession of an algorithm that is capable of gauging audience characteristics and offering the most attention-attracting product becomes the main determinant of market success, sparking competition to provide the most enthralling incarnation of algorithmic governance of attention.

In effect, attention is a real commodity that is extracted, sold, and bought (Sunstein 2018: 229). And as predicted by the basic economic model of supply and demand, as the demand for attention (the number of actors competing for audience attention) increases and the supply of it fails to catch up (since we are by nature limited in our capacity to consciously engage with the environment), scarcity ensues, driving the price upwards; moreover, as attention thereby becomes ever more valued and sought for, products and services claiming to be able to catch, retain, and harvest attention (typically with the help of allegedly ever more precise employment of data) proliferate, in the end only further increasing attention scarcity (Léveillé Gauvin 2018: 293). Since everybody understands that attention must be grabbed in order to affect human behaviour, the competitors must equally be conscious that many other actors are aiming to grab that same unit of attention simultaneously, further precipitating the enthrallment race (Vaidhyanathan 2018: 80). And as the idea of multitasking is being increasingly put in question by neuroscientific research (Romaniuk and Nguyen 2017: 911), it seems that even the amount of attention available on offer might be lower than previously anticipated. Hence, governance of attention, either through ex ante knowledge of likes and interests or through tampering with affective flows is highly sought after.

What strikes in the above is the mundanity of algorithmic governance. While laws and regulations in their traditional sense mainly deal with the key interactions, provisions, and dispossessions of life, algorithmic governance, based on the digital representation of almost the entirety of life, affects the detail of everyday experience and allocation of attention (filtering of what is to be experienced) by not only passively constituting and structuring the digital environment but also—and perhaps even more importantly—doing that interactively, responding to and simultaneously conditioning the ways in which the environment is encountered and reacted to by individuals. Just like we are by now already used to targeted advertising following us and responding to our web use patterns, we should get used to the idea of the governmental affordances of code, which is tracking, reacting to, and affecting us at every point of our digital life (provided there is still legitimate ground for separating digital and non-digital life).

Moreover, since interactivity means that everybody ultimately becomes wrapped in an (at least somewhat) different tightly fitting version of code-driven normative architecture, the non-transparency of algorithmic governance is only further increased. This crucial affordance also allows actors to 'to zero in on those believed to be the most receptive and pivotal audiences for very specific messages while also helping to minimize the risk of political blowback by limiting their visibility to those who might react negatively' (Nadler et al. 2018: 7). Moreover, this differentiation allows for even further fine-tuning through the capacity to test different variants of the same ad and different parameters for targeting, thereby achieving ever-greater engagement optimisation (Nadler et al. 2018: 7; see also Bedingfield 2019; Lomas 2019). The effect is rather clear: although the promise of (micro)targeting is 'making advertisements more relevant to users', it is with great ease that the same data-driven techniques are being employed 'to make users more pliable for advertisers' (Nadler et al. 2018: 5). Therefore, the ultimate end result is a shift of perception towards users as 'programmable objects' that are enthralled and governed 'through hyper-personalized technologies that are attuned to our personal histories, present behaviors and feelings, and predicted futures' (Frischmann and Selinger 2018: 9–10). And, since algorithmic governance typically does not manifest itself extrinsically but acts through emotional triggering, it is much more difficult to notice (but also, because it plays on pleasure, we may well simply not be bothered to notice it as long as it feels good). In the latter sense, algorithmic governance is a bit like the Heideggerian nature of a tool: we only become aware of its nature and function when it is broken or does not work properly anymore (e.g. when the targeting goes awry), giving us a glimpse of Being (see Heidegger 1996).

The preceding analysis might, at some stages, perhaps invoke reminiscences of actors like Cambridge Analytica or AggregateIQ. Nevertheless, due to the high profile revelations pertaining to the activities of such actors and the corresponding amount of public attention paid to them, this book aims at going beyond narrow focus on specific rogue data-wielding actors, not least because the actual impact of each of their activities, which are usually campaign-focused, is difficult (if not impossible) to measure. Instead, in order to fully appreciate the scope and impact of algorithmic governance, one must focus on the mundane, i.e. the everyday (in fact, minute-to-minute) influence exerted by data-based algorithmic sorting, herding, and ranking, the architectural affordances of the online environment, and the enthralling pleasure of data-based nudge.

4.2 THE ALGORITHMIC PROMISE OF NUDGE

It is important to keep in mind that individual decision-making rarely takes place on a rational, emotion-neutral terrain: Instead, 'much individual decision-making occurs subconsciously, passively, and unreflectively rather than through active, conscious deliberation' (Yeung 2017: 120; see also Kahneman 2013). Indeed, the decisions that humans make in actual real-life circumstances are typically 'decisions they would not have made if they had paid full attention and possessed complete information, unlimited cognitive abilities, and complete self-control' (Thaler and Sunstein 2009: 5–6). As a result, algorithmic systems capable of sorting and ranking choice options available to individuals and then continuously learning from immediate reactions and continuous behaviour of such individuals in order to re-alter the landscape that individuals find themselves immersed in become particularly potent governance mechanisms. Such mechanisms do not openly force people to choose a particular option but, instead, shape the choice environment (in terms of available information, understanding of the world, and accessible solutions) in such a way that only the preset option (as in purchase or voting choices) remains likely to be taken.

Certainly it must be admitted that such governance through choice architecture, including—and perhaps especially—the algorithmic one, is ambitious: after all, developing 'a technical blueprint for a desired reality' is complicated by the unpredictability not only of the environment but also of human action, since users can easily start interacting with the architecture in unanticipated ways if the latter give them more utility than ones intended by the developer (Kleve and De Mulder 2005: 318). Nevertheless, the role of the data input is precisely to minimise the likelihood of such architectural tinkering going awry. And if users do not even want to *think about* alternatives because the most pleasure-inducing options are stacked in front of them, then pre-programmed outcomes become significantly less difficult to achieve.

As already noted by Lessig, for the code writers, the architecture-creating and access-managing capacity of their trade is bound to become 'a means to achieving the behaviors that benefit them best' (Lessig 2006: 84). Effectively, algorithmic analysis of data enables our cyber-physical environment to intervene in order to either pre-empt undesirable behaviours or to encourage desired ones, if not due to a deep knowledge of our thoughts and intentions (although, arguably, such knowledge is not impossible) then at least through an acquired understanding of our behavioural patterns (Hildebrandt 2016: 29). Moreover, this shaping continuously unfolds in real time, dynamically shifting in reaction to any changes in behaviour or emotional states (Yeung 2017: 122). Hence, we become co-constitutive of the environment through our use of it but simultaneously that co-constitution happens through no effort or intention of ours but under the conditions set through the means provided by the code-wielding architects of algorithmic governance.

After all, the proper way of understanding human choices involves focusing on the *perceptions*, rather than substance, of outcomes that are, more often than not, dependent on otherwise supposedly irrelevant factors (Thaler 2015: 4–5). Such factors may include the placement of objects, architectural design (including in terms of code-as-law-as-architecture), pre-allocation of attributes and possessions, previous investment (including emotional), or just simple play on informational affordances. In other words, people are typically neither capable nor in a position to make optimal decisions (or even something approximating them), at least in a majority of cases, even though 'we all enjoy having the right to choose for ourselves, even if we sometimes make mistakes' (Thaler 2015: 324). It is in this context that actors wielding algorithmically derived knowledge are empowered to nudge us in a predefined direction that suits their interests. And while this nudging can often be political or economic (e.g. towards a particular voting or purchase behaviour), it also assumes more personal and intimate forms, such as nudging in the sense of altering the behaviour of the quantified self (Thomas et al. 2018: 9).

As a result, a particularly important role is played by the so-called choice architects. According to Thaler and Sunstein (2009: 3), a choice architect is somebody in a position that renders them capable of 'organizing the context in which people make decisions', such as arranging the offerings in a particular order that makes one choice more likely than the others or setting a particular option as the default one and thus requiring individuals to put in some effort (e.g. opt out or change) if they are not to go with it. In slightly cruder terms, Nadler and McGuigan (2018: 153) emphasise that in practice the application of behavioural approaches leads to the prospect of shaping behaviour by 'manipulating the contexts in which decisions are made'. Unsurprisingly, therefore, marketers working in various contexts (business, political, etc.) 'seek control over choice architecture to promote their clients' interests' (Nadler and McGuigan 2018: 153). The idea is, in essence, to still allow people to choose but to make them do this not under the conditions of their own making.

In an algorithmic context, choice architects are even more powerful: They not only rearrange reality by making some elements more prominent than others—they *create* reality altogether by adding and eliminating certain choices as such as well as particular conditions of choosing. One only has to think of the design, built-in (and left out) functionalities, and other affordances of various platforms to understand the scale of both empowerment and the constraints. On other occasions, the very threat of invisibility, such as on social media or among search results, is enough to strongly nudge actors towards pre-programmed decisions, with platform owners not infrequently manipulating their users by disclosing elements of algorithmic ordering to some users but not to others (Cotter 2018: 4; see also Bucher 2018: 73). Hence, the algorithmic architecture of code, discussed in the previous chapter, can in itself also be conceived of as a *choice architecture*. In its least intrusive, it can act identically to the offline one—through a nudge—but in its most intrusive, it can simply eliminate all unwanted choice options or entire choice scenarios. After all, it must be remembered that the digital landscape has no natural constraints and is, therefore, open to any architectural (re)configurations. In fact, more critically minded authors, such as Frischmann and Selinger (2018) would straightforwardly equate such choice modelling with sociotechnical engineering.

Very often, there is even no need to substantially manipulate humans on some deep level—it is only enough to nudge them in the required direction, particularly when they are unsure about what is to be done next.

For example, it is natural for humans to be unsure—they lack certainty on many complex issues, and in such cases they typically gravitate towards the middle ground because that is the terrain most accommodating of tentative opinions; nevertheless, once they find agreement with their tentative views or when there appears to be a tendency in their environment to favour a particular option, people become more convinced, and the more confirmation they encounter, the more confident and extreme they become (Sunstein 2018: 74). As a result, in such cases, it is sufficient if one is capable of bending (or stacking in advance) the algorithmic architecture in a way that puts the undecided together with the convinced in order to cause a substantial change in thinking.

The employment of data in order to predict, target, and change human behaviours and emotions benefits from both the ability to 'hook' individuals as well as from behavioural scientists' ideas about triggering within individuals or nudging them towards particular responses and decisions (Williamson 2017: 271). Such big data-driven processes of nudging (or, rather, 'hypernudging') individuals towards preset goals are particularly 'nimble, unobtrusive and highly potent' not only because they are much less noticeable than other, blunter and more openly forceful, techniques of governance and control but also due to their highly personalised nature—the choice environments in which individuals become immersed can easily be individually tailored (and morph *with* the individual), making it excessively difficult *not to* submit to the nudge (Yeung 2017: 122, 130). Indeed, this big data-based hypernudging (sometimes also alternatively referred to as 'big nudging') can easily be seen as large-scale coordination at best and large-scale manipulation at worst (see Helbing et al. 2017).

To reiterate, users are being profiled, their cognitive biases uncovered and the most susceptible triggers identified in order to allocate tailored persuasion strategies and thereby maximise the likelihood of the target being convinced (Nadler and McGuigan 2018: 153; see also Calo 2014), making it possible 'to persuade users and forge new behaviors' through the process of 'measuring and applying various behavioral data' (Shin and Kim 2018: 168). The end result is, therefore, 'a frictionless world that surreptitiously adjusts the environment to the needs and desires of its users' (Hildebrandt 2016: 4). In a sense, individuals collude with data-rich actors: the former open themselves to being 'continuously, pervasively and increasingly subjected to Big Data hypernudging strategies' in return for a bespoke and continuously optimised environment that seemingly 'knows' how to shape itself to be as consumer satisfaction-maximising as possible; nevertheless, since in opening themselves in such a way individuals rescind most of the actual agency that they have (the agency that remains is illusionary only, effectively leaving only one option available), such collusion is highly asymmetric (Yeung 2017: 131).

A further issue of code-as-law-as-architecture is not only that it can be used to nudge individuals towards preselected goals but also that such nudging can easily find its way into traditional law as well. As the likelihood of public demand for laws and regulations on a specific issue is, effectively, a function of the availability to mind of a problem, its perceived frequency of manifestation, and its salience (Jolls et al. 1998: 1518–1519), algorithmic control of choice architecture as well as the knowledge of what triggers have to be stimulated to move humans in necessary directions becomes a way of managing the demands for and expected content of conventional law. Again, this could serve as a further iteration of the increasing futility of separating between the online and the offline worlds as well as between their modes of governance.

Notably, in order for algorithmic governance to be successful, it is not even necessary to build a completely 'real' representation of the person or a group—it is sufficient to be able to determine what is necessary to nudge them towards a desired action. As such, the 'libertarian paternalism' of nudge 'tries to influence choices in a way that will make choosers better off, *as judged by themselves*' (Thaler and Sunstein 2009: 5), ostensibly helping to achieve goals without limiting choices (Thaler 2015: 324). Nevertheless, the obvious question here is whether we are really in control of the conditions and the substance of our judgement, particularly in the digital environment, due to the latter being malleable and structured in accordance with corporate interest. After all, judgements are typically motivated by certain inputs (the conditions of choosing) and the presence of sought outcomes (goals to be achieved). And if both the inputs and the outcomes available for expectation are shaped by the architecture of code, it is by no means impossible to redesign that architecture in a way more aligned with preset outcomes. In this case, criteria for even a rational judgement should be seen as dependent upon the interests of those in charge of code (either the writers or those commissioning it to be written). Code is, after all, law, and any changes in the regulatory environment also change the ranking of outcomes by reassigning costs and benefits and thereby making one choose not necessarily the most outcome-maximising option overall but only one that is the most outcome-maximising within a given regulatory environment.

Overall, then, of prime importance is the power that choice architects have to exploit the logic of 'libertarian paternalism' by arranging the choice environment in order to advance pre-programmed goals; However, due to the private nature of such regulation and the opacity of choice-structuring algorithms, it is impossible to verify whose interest is being advanced in this way. Thaler and Sunstein (2009: 239) readily admit that misuse of nudging for private gain is a possibility worth worrying about; however, according to them, solutions can be found: rules upholding competition, reducing fraud and the power of interest groups, and creating incentives to serve the public interest should be created, but even more importantly, 'a primary goal should be to increase transparency'. It is precisely this point that is the most problematic in case of algorithmic governance: as algorithms are non-transparent due to their nature and legal and commercial status, it is difficult to imagine accountability thus stipulated. If the code that builds architecture, determines informational affordances, and sets causal chains cannot be examined and its goals assessed, then trust becomes difficult to achieve. Moreover, since code is a private business matter, even the expectation of it serving the public interest would seem misplaced. And while algorithmic governance based on some sort of public-private partnership is not inconceivable, it is the predominantly private incarnation that is generating the experience and the affordances of the digital environment, as evidenced by the most widely used platforms and services.

Finally, it has to be kept in mind that the whole edifice of nudging is constructed around the distinction between what Thaler and Sunstein (2009) call the Humans and the Econs. The Humans are the fallible beings displaying only bounded rationality and willpower (see also Jolls et al. 1998: 1477–1479) that are encountered throughout this book. The Econs, by contrast, are the ideal rational utility maximisers stipulated by classical economic theory. The trick is, therefore, to allow the Humans to retain their humanity but enable (or perhaps even make) them choose as wisely the Econs would. It is the function of a nudge to do that trick, the premise being that that of good Econs determining what the right and rational decision is and then choosing the correct nudge. However, bad Econs (Becons) and indifferent Econs (Iecons) should also be presumed to exist. Becons, whose aim is to deliberately nudge Humans towards what they (being rational) know to be the wrong choice are less interesting for the purpose of this analysis. Iecons, however, should be given more consideration. Iecons know what is best for *them* and strive for that, nudging Humans to make Iecon-return-maximising choices along the way but being indifferent about whether such choices are good for the Humans or not. The very structure of datafied algorithmic governance, discussed in the previous chapter, opens the gates wide for Iecons. And since 'a great majority of nudges exploit cognitive biases, rather than actually attempting to expand the capability and the commitment of individuals and their social contacts to invest in learning' (Gandy and Nemorin 2018: 12), it is more than likely that Iecon-return-maximisation will prevail long-term. Which, in turn, leads us to interrogating the very issue of agency in choice.

References

Balazs, J. A., & Velásquez, J. D. (2016). Opinion Mining and Information Fusion: A Survey. *Information Fusion, 27,* 95–110.

Bedingfield, W. (2019, August 5). Boris Johnson's Facebook Advert Splurge Is All About Data. *Wired.* Available at https://www.wired.co.uk/article/conservative-boris-johnson-facebook. Accessed 6 Aug 2019.

Brigham, T. J. (2017). Merging Technology and Emotions: Introduction to Affective Computing. *Medical Reference Services Quarterly, 36*(4), 399–407.

Bucher, T. (2018). *If… Then: Algorithmic Power and Politics.* Oxford and New York: Oxford University Press.

Calo, R. (2014). Digital Market Manipulation. *The George Washington Law Review, 82*(4), 995–1051.

Citton, Y. (2017). *The Ecology of Attention.* Cambridge and Malden: Polity Press.

Colvin, S., & Kingston, W. (2017, July 7). *Why Conversation Is the Future of Customer Experience.* PwC. Available at https://www.digitalpulse.pwc.com.au/conversation-customer-experience-cco-study. Accessed 23 June 2018.

Cotter, K. (2018). Playing the Visibility Game: How Digital Influencers and Algorithms Negotiate Influence on Instagram. *New Media & Society.* https://doi.org/10.1177/1461444818815684.

Dahlgren, P. (2018). Media, Knowledge and Trust: The Deepening Epistemic Crisis of Democracy. *Javnost—The Public, 25*(1), 20–27.

Damasio, A. (2018). *The Strange Order of Things: Life, Feeling, and the Making of Cultures.* New York: Pantheon Books.

Etter, M., et al. (2018). Measuring Organizational Legitimacy in Social Media: Assessing Citizens' Judgments with Sentiment Analysis. *Business and Society, 57*(1), 60–97.

Evans, E. (2016). The Economics of Free: Freemium Games, Branding and the Impatience Economy. *Convergence: The International Journal of Research into New Media Technologies, 22*(6), 563–580.

Frischmann, B., & Selinger, E. (2018). *Re-Engineering Humanity*. Cambridge and New York: Cambridge University Press.

Gandy, O. H., & Nemorin, S. (2018). Toward a Political Economy of Nudge: Smart City Variations. *Information, Communication & Society*. https://doi.org/10.1080/1369118X.2018.1477969.

Giatsoglou, M., et al. (2017). Sentiment Analysis Leveraging Emotions and Word Embeddings. *Expert Systems with Applications, 69*, 214–224.

Heidegger, M. (1996). *Being and Time*. Albany: State University of New York Press.

Helbing, D., et al. (2017, February 25). Will Democracy Survive Big Data and Artificial Intelligence? *Scientific American*. Available at https://www.scientificamerican.com/article/will-democracy-survive-big-data-and-artificial-intelligence/.

Hildebrandt, M. (2016). Law *as* Information in the Era of Data-Driven Agency. *The Modern Law Review, 79*(1), 1–30.

Jolls, C., Sunstein, C. R., & Thaler, R. (1998). A Behavioral Approach to Law and Economics. *Stanford Law Review, 50*, 1471–1550.

Just, N., & Latzer, M. (2017). Governance by Algorithms: Reality Construction by Algorithmic Selection in the Internet. *Media, Culture and Society, 39*(2), 238–258.

Kahneman, D. (2013). *Thinking, Fast and Slow*. New York: Farrar, Straus and Giroux.

Kalpokas, I. (2019). *A Political Theory of Post-Truth*. London and New York: Palgrave Macmillan.

Kleve, P., & De Mulder, R. (2005). Code Is Murphy's Law. *International Review of Law, Computers & Technology, 19*(3), 317–328.

Klinger, U., & Swensson, J. (2018). The End of Media Logics? On Algorithms and Agency. *New Media & Society*. https://doi.org/10.1177/1461444818779750.

Lessig, L. (2006). *Code: Version 2.0*. New York: Basic Books.

Léveillé Gauvin, H. (2018). Drawing Listener Attention in Popular Music: Testing Five Musical Features Arising from the Theory of Attention Economy. *Musicae Scientiae, 22*(3), 291–304.

Lomas, N. (2019, August 6). *UK Watchdog Eyeing PM Boris Johnson's Facebook Ads Data Grab*. Tech Crunch. Available at https://techcrunch.com/2019/08/05/uk-watchdog-eyeing-pm-boris-johnsons-facebook-ads-data-grab/. Accessed 6 Aug 2019.

Nadler, A., Crain, M., & Donovan, J. (2018). *Weaponizing the Digital Influence Machine*. Data & Society Research Institute. Available at https://datasociety.

net/wp-content/uploads/2018/10/DS_Digital_Influence_Machine.pdf. Accessed 6 Aug 2019.

Nadler, A., & McGuigan, L. (2018). An Impulse to Exploit: The Behavioral Turn in Data-Driven Advertising. *Critical Studies in Media Communication, 35*(2), 151–165.

Papakyriakopoulos, O., et al. (2018). Social Media and Microtargeting: Political Data Processing and the Consequences for Germany. *Big Data & Society.* https://doi.org/10.1177/2053951718811844.

Poria, S., et al. (2017). A Review of Affective Computing: From Unimodal Analysis to Multimodal Fusion. *Information Fusion, 37,* 98–125.

Puschmann, C., & Powell, A. (2018). Turning Words into Consumer Preferences: How Sentiment Analysis Is Framed in Research and the News Media. *Social Media + Society.* https://doi.org/10.1177/2056305118797724.

Riccio, J. (2017, August 22). *Why the Experience Age Is Closing the Gap Between Consultancy and Agency.* PwC. Available at https://www.digitalpulse.pwc.com.au/experience-age-advertising-agency-consultancy. Accessed 23 June 2018.

Romaniuk, J., & Nguyen, C. (2017). Is Consumer Psychology Research Ready for Today's Attention Economy? *Journal of Marketing Management, 33*(11–12), 909–916.

Schwark, J. D. (2015). Toward a Taxonomy of Affective Computing. *International Journal of Human-Computer Interaction, 31,* 761–768.

Serrano-Guerrero, J., et al. (2015). Sentiment Analysis: A Review and Comparative Analysis of Web Services. *Information Sciences, 311,* 18–38.

Shin, Y., & Kim, J. (2018). Data-Centered Persuasion: Nudging User's Personal Behavior and Designing Social Innovation. *Computers in Human Behavior, 80,* 168–178.

Sunstein, C. R. (2018). *#Republic.* Princeton and Oxford: Princeton University Press.

Thaler, R. H. (2015). *Misbehaving: The Making of Behavioural Economics.* London and New York: Penguin Books.

Thaler, R. H., & Sunstein, C. R. (2009). *Nudge: Improving Decisions About Health, Wealth and Happiness.* London and New York: Penguin.

Thomas, S. L., Nafus, D., & Sherman, J. (2018). Algorithms as Fetish: Faith and Possibility in Algorithmic Work. *Big Data & Society.* https://doi.org/10.1177/2053951717751552.

Vaidhyanathan, S. (2018). *Anti-Social Media: How Facebook Disconnects Us and Undermines Democracy.* Oxford and New York: Oxford University Press.

Williamson, B. (2017). Moulding Student Emotions Through Computational Psychology: Affective Learning Technologies and Algorithmic Governance. *Education Media International, 54*(4), 267–288.

Yeung, K. (2017). 'Hypernudge': Big Data as a Mode of Regulation by Design. *Information, Communication & Society, 20*(1), 118–136.

CHAPTER 5

Agency and the Posthuman Shape of Law

Abstract The changes and transformations described in the previous chapters necessitate a reconsideration of human agency. However, it is important not to jump to conclusions: whereas it is clear that accounts of human privilege in agency are no longer sustainable, algorithms equally cannot be seen as the unconditional masters of (human) life. Instead, agency is demonstrated to be located in assemblages composed of humans, code, and technological artefacts that integrate and shape the contours of everyday life. In order to appropriately conceptualise the matter, this chapter turns to posthumanist thought, particularly its emphasis on the relationality and embeddedness of human existence. The outlook thus developed rejects the longstanding dominance of anthropocentrism and allows for multiple ways of treating nonhuman agents on par with human ones.

Keywords Posthumanism · Agency · Assemblage · Code · Causation · Choice

The increasingly prevalent role of algorithms in the regulation and determination of human action raises prominent questions regarding the role of free will and, particularly, agency. In other words, it is no longer clear who the active agent making the choice and forming an opinion is: the human person or computer code. And while some of the more radical versions that accord to algorithms an almost unlimited power at the

© The Author(s) 2019   67
I. Kalpokas, *Algorithmic Governance*,
https://doi.org/10.1007/978-3-030-31922-9_5

expense of humans arguably go too far, it is indeed the case that humans have become increasingly dependent upon their creations. Nevertheless, the exact location of agency remains unclear: on the one hand, humans have certainly lost their privileged position as their decisions and actions are effectively determined by code (because the circumstances are stacked so that it is impossible to choose otherwise), and even the writers of code are rendered dependent upon the results of algorithmic analysis of data for the construction and refinement of their algorithms. At the same time, algorithms are social creations dependent on their writers and on the reactions of their users that inform their own future iterations. Hence, one should better think of relational assemblages within which agency is dissolved. Still, because agency can no longer be seen as a fully human domain, posthumanist thinking becomes especially acute. Indeed, the key tenet of this line of thinking—that there is nothing exclusive about being human and humans are merely one part of a much greater picture—should be seen as the most important explanatory framework. As a result, this chapter concludes with a consideration of algorithmic governance in the light of technological posthumanism.

5.1   The Distributed Assemblage of Agency

Algorithms must increasingly be seen as 'actors and policy-makers' on their own right (Just and Latzer 2017: 254). As a result, the traditional focus on agency and independent standing as an exclusively human phenomenon is becoming ever less adequate: Algorithms are progressively acquiring agency of their own, standing on par with and even, it would not be unreasonable to claim, in certain contexts exceeding the power and influence of humans and becoming key influencers of the latter. As stressed by Schwitzgebel and Garza (2015: 100), it is immaterial what kind of physical or mental architecture one possesses (digital-mechanical or biological)—or, perhaps, whether one *has* a physical architecture, i.e. a material body as such—insofar as one is capable of affecting and influencing somebody else's social and psychological characteristics. And as digital artefacts become capable of making largely independent decisions that are highly consequential, their agentive status should be beyond serious doubt (Vladeck 2014: 121). Digital artefacts, particularly those endowed with a machine learning capacity or those that can be ascribed to the category of artificial intelligence, differ from simple tools, such as a hammer, precisely due to their own quality of agency and interactivity in watching, responding, organising, guiding, and entertaining our lives among many other things (Mahon 2018: 80) in ways that exceed mere predefined pathways from inputs to outputs, from an 'if' to a 'then' but are, instead, multiple and largely unpredictable (Bucher 2018: 23–24).

The progressive use of data and algorithms to both analyse the former and act upon it establishes a specific human–algorithm relationship that, almost by definition, already implies some sort of human subjection to or, at the very least, dependence upon, algorithms (Lyon 2014: 6). At the same time, Mayer-Schönberger (2008) strongly criticises Lessig (2006) for assuming a strong unidirectional relationship through which technology determines society—and quite rightly so. Indeed, the role of society should not be overlooked in discussions of human and algorithmic agency, not least because the data input that shapes the goals of the developers and wielders of algorithms, and then often directly algorithms themselves through numerous feedback loops, comes from the society itself. In addition, it must be kept in mind that 'algorithms are bits of code that only have meaning when deployed within a specific context' and, therefore, are not (yet) endowed with the full autonomy of isolated deterministic actors—instead, according to this argument, they are better conceived as 'an inextricable component within a network of communicative practices' (Carlson 2018: 1761). Under this line of argument, algorithms lack proper agency because they can only do (or, in case of machine learning algorithms, acquire the capacity to do) what humans have set for them (Klinger and Swensson 2018).

In the above sense, algorithms may still be seen to retain a tool-like nature. In part that is correct: They are written by humans with specific tasks and purposes in mind. The issue, however, is with the *means* for achieving such ends: once deployed, the algorithmic agents become co-constitutive of the environment in which they operate, not only regulating but also more subtly shaping human action (and, as shown in the previous chapter, also affective flows and patterns of thought) through the architecture that they create. As Bucher (2018: 72–73) accurately stresses 'algorithms have productive capacities not merely by mediating the world' but also by working upon the world in which they are deployed and 'by making a difference in how social formations and relations are formed' (Bucher 2018: 72–73). As such, they can clearly be considered to be active driving forces within our world.

Still, despite (or perhaps *because of*) the 'distinct algorithmic moment in contemporary zeitgeist' that results in the more complicated or intentionally black-boxed algorithms 'evoking in us the feelings of a technological sublime in all its awe-inspiring, rationality-subsuming glory' (Ames 2018: 1–2), some caution with regards to algorithmic agency is necessary. After all, the role that algorithms play is part and parcel of the broader cultural processes and, therefore, 'algorithms have everything to do with the people who define and deploy them, and the institutions and power relations in which they are embedded' (Ames 2018: 3; see also Thomas et al. 2018: 9–10). The issue is perhaps most explicitly expressed by Schraube (2009: 305): 'the action of things does not emerge from a void. However mediated, its origins lie in human action and, in turn, the action of things effects human action'. Others, meanwhile, go even further in implicating humans. For example, Kelleher and Tierney (2018: 33–34) stress that, at least for now, human data analysts are still needed to prepare ground for, and to work alongside, algorithms, as opposed to the popular myth of algorithms simply being let loose on data, especially when the analytical pattern and information discovering function is performed (perhaps less so with automated sorting, ranking, and retrieval); hence, the argument goes, algorithmic governance is not (yet) self-contained.

Nevertheless, even for the analysts as well as the writers and the wielders of code, their dependence on return data and algorithmic feedback loops for refinement of products, campaign offerings, etc. implies that the creator becomes dependent on the created. At the very least, one can clearly observe a relation of co-production in this context. In an important way, though, the relationship between humans and algorithms is reminiscent of the Hegelian master–slave dialectic, in which the master, although nominally superior, is in fact the weaker party whose identity is not independent but constituted through recourse to the nominally inferior slave and impossible without the latter (see Hegel 2018). Hence, while situation is hardly as dramatic as algorithmically predetermined decisions leading to 'to a brainwashing of the people', reducing the latter to 'mere receivers of commands, who automatically respond to stimuli' (Helbing et al. 2017), it is nevertheless undoubtedly the case that conditioning by and dependence upon algorithms is a defining factor in today's societies.

Still, artefacts, such as those based on intelligent code, do not merely have consequences instead of agency, as Hornborg (2017) claims—they have interactive capacity and are capable of dynamically changing our choice environment by learning from the past, analysing and affecting the present, and foreseeing the future. It must be noted that our current life-world itself 'thrives on mobile, hyper-connected cybernetic systems such as smartphones, online social media, gaming, search engines, health and fitness apps, fraud detection systems and the more' (Hildebrandt 2016: 5), clearly inferring that the very essence of our technologically enabled (not merely enhanced) existence is rendered contingent upon the performance and the enabling role of code and of devices enabled by code. Definitely, it is important not to surrender to some kind of fatalism. Here, Klowait (2019: 13) is correct in asserting that the reconceptualization of non-human technological artefacts as on par with humans must not result in, for example, naturalisation of inequalities produced and/or expressed by and through such artefacts (see also Schraube 2009: 305).

As noted by Hildebrandt (2016: 5), this world has a double presence: It 'has a frontend (the world we see and navigate) and a backend (the largely invisible computational architecture that sustains and informs the frontend)'. By running the backend, therefore, algorithms make decisions about value, relevance, and visibility, establishing and allocating privilege (Cotter 2018: 4). Hence, again, the lived and perceived environment is sustained by an architecture of hardware and code that collectively ensure the seamless, personalised, and, to the largest degree possible, pleasurable life. In this context, the relationship between humans and algorithms can be strongly ambiguous. On the one hand, algorithms sit in-between two groups of humans: the designers and the users (Janssen and Kuk 2016: 374). On the other hand, though, the two roles are fluid: even if we agree that designers have a higher level of agency vis-à-vis the users, that is only with regards to a single algorithm in question. After all, someone who is the designer of one algorithm is the user (and perhaps even a subject) of another, for example with the aim of 'hacking' their life for greater efficiency (see e.g. Reagle 2019), as well as of many others. Increasingly, therefore, we inhabit a digitised world that seems to possess agency of its own, morphing and evolving as it interacts with us (on a par with *our* interactions with this world), thus being engaged in a loop of co-constitution—co-constitution of both the world and of us by us and by human-made artefacts.

It should by now be clear from the preceding that we are witnessing our 'increasing dependence on technologization processes and their complete penetration of contemporary culture' (Herbrechter 2013: 81). That is witnessed, in particular, in the 'transformations of identity and subjectivity, as well as changes and extensions of experience, which go far beyond the traditional individual liberal humanist understanding of the subject' as well as a wide range of possibilities and potentialities that range from 'new forms of surveillance, state and increasingly private and corporate biopolitics, processes of deindividualization and dehumanization' to 'improved quality of life and life expectancy, as well as new forms of electronic social interaction' (Herbrechter 2013: 191). In this environment, humans have to either rely upon or 'game' the algorithms in order to achieve their aims (see e.g. Cotter 2018), in either case underscoring the key role played by code.

Notably, in order to properly analyse power and politics, one must not only understand 'the way in which some realities are always strengthened while others are weakened' but also (and that is an important qualitative difference from the past) 'to recognize the vital role of non-humans in co-creating these ways of being in the world' (Bucher 2018: 3). Moreover, to an extent such influence can be exercised by non-human elements of the system even regardless of their own qualities but through their sheer presence, not least because 'knowing algorithms might involve other forms of registers besides code' in the sense that the conscious effects on human action (actions human users *choose* to take or not to take, as opposed to being nudged or facing a straightforward inability to do certain things) depend on how the attributes of algorithms are *perceived*, i.e. 'articulated, experienced and contested in the public domain' (Bucher 2017: 40). This multilayered displacement of human action and choice, in turn, calls into question the standard assumptions about individual autonomy and the role of humans as the paramount shapers of the world, both social and physical.

Moreover, as the main danger that we, as individuals, face 'shifts from privacy to probability', with algorithms predicting the likelihood of health conditions, capacity to repay a loan, propensity to engage in criminal activities, etc., both human autonomy and free will are, if not nullified, then at least ignored: humans are reduced to a set of algorithmic if-thens whereby datafied attributes lead to associated outcomes as if humans had no agency whatsoever, effectively punishing people in advance or, at the very least, focusing on their propensities rather than actions (Mayer-Schönberger and Cukier 2017: 17, 170; see also Beer 2019: 4). As a result, 'a guarantee of human agency' that would guard humans 'against the threat of a dictatorship of data' might be needed in order to protect human freedom of choice and meaningful responsibility for their actions (Mayer-Schönberger and Cukier 2017: 176–177). Moreover, as decisions and predictions are increasingly being made based on data that are too expansive to comprehend and by algorithms that are unknowable, we are facing a situation in which it might no longer be possible to explain the decisions that we make or that we are subjected to (Mayer-Schönberger and Cukier 2017: 17). In this context, calls for algorithmicising the judicial process (potentially up to installation of robojudges—see e.g. Tegmark 2017: 105–106) represents a further step towards subjection of humans to abstract computational models.

Real agency should, then, be seen to reside in assemblages of 'human actors, code, software and algorithms' that shape how existing data circulate and how new data are being created based on the present data and their circulations (Beer 2017: 476). Indeed, as already noted above, algorithms neither originate ex nihilo nor operate in a humanless void— they are 'inherently relational and begin with human endeavor and operate with the knowledge of the people to whom they serve' and should be seen as part of a system involving 'an intricate, dynamic arrangement of people and code', the output of which is typically difficult to comprehend and predict even for the insiders (Janssen and Kuk 2016: 374). There is constant dynamic within such agglomerations: as Lupton (2018: 3) suggests, digital assemblages that, for her, 'bring together humans-devices-software-data-space-time' are best conceived of as 'lively'. Due to their relational nature, such assemblages 'cannot merely be reduced to [their] constituent parts', leaving the human subject permanently entangled with 'agential capacities of nonhumans' and meaning that ultimately agency is *distributed* (Bucher 2018: 50–51). Here we have to deal with an impersonal infrastructural subjectivation, which 'establishes a set of relationships that mobilize and aggregate users and non-users with non-human data points' without being 'constrained or otherwise limited to a person' any longer (Langlois and Elmer 2018: 3).

Notably, then, 'current impersonal subjectivation processes do not focus solely on the person, but on orchestrating a set of relations among groups, humans, non-humans, services or products, places, spaces, technologies, and times' (Langlois and Elmer 2018: 11), producing assemblages rather than hierarchies or clearly defined causal chains. In effect, the self as such is put into question: If one cannot lead an independent life but, instead, one's experiences, affects, cognition, and choices originate, at least in part, outside the human body and can even be deliberately induced in order to nudge the individual towards a predestined outcome, then the self, as a conscious cognitive entity, becomes disentangled from a particular body. Instead, the self starts being best conceivable as an assemblage itself, i.e. 'a distributed, networked self that constantly emerges at various intersections between humans, non-humans, objects, materials and energy flows' (Pötzsch 2018: 3314). Although the language of the preceding quote might, at first sight, even seem to point towards some sort of mysticism, nothing of the sort is implied. At stake here is the digital-technological extension (and, due to the loss of agency, simultaneously also reduction) of the self.

Causation, even of that of which the person themselves is an apparent cause, is also rendered problematic by the above: due to aforementioned impersonality, one may not necessarily become a cause due to intentional action but through algorithmically interpreted aggregations of the most mundane life data. Consequently, regulation becomes a distributed assemblage itself, composed of traditional human-made law, human agency, the architecture of code, and the effects of that code in action. Crucially, algorithmic governance shapes not only human action but also the will that is written into subsequent iterations of code and is constantly renegotiated within every relational assemblage.

This turn towards 'de-privileging the human' has broad-ranging consequences: The self ceases to be a universal point of reference, the singular locus of *cogito*, and instead re-emerges as 'just another object with certain agentic capacities' while identities, both individual and collective, 'become conceivable as fluid, hybrid and constantly evolving' (Pötzsch 2018: 3314). In fact, the self as such becomes hardly conceivable as singular, with data repositories containing and springing into life one's doubles, or doppelgangers, that obtain a real-life existence of their own through information feedback loops. In effect, there are versions of us residing on the platforms and databases of service providers and institutions that, obviously, relate to our actual selves that they are derived from but, nevertheless, are just a matter of data captured and analysed, springing from our use of that specific bit of the overall algorithmic architecture of the digital world. Of course, as noted earlier, the ultimate goal would be that of integration of those varied selves—that is what the operating system of entire life, predicted by Vaidhyanathan (2018), would do. That, nevertheless, remains a future possibility. At the moment, meanwhile, we do not yet experience algorithmic governance as a single totality but as a distributed system that engages with us slightly differently depending on what kinds of doppelgangers reside in places that we go to.

The preceding should not be read as a claim that algorithmic governance is weaker than expected—it is merely more variable. Moreover, it also is a source of greater fluidity and variability of the self. If there was just a single totality of algorithmic governance, a single version of the self would be moulded across platforms, services, and institutions in accordance with the wishes of some master strategist, either human or algorithmic, wielding power that even Big Brother would have only dreamt of. In the current world, where the self, through merely existing, becomes the source and inspiration for multiple doppelgangers, there is, of course, a reciprocal relationship because such doppelgangers, in return, become sources of inspiration for algorithmic sorting and nudging practices. Those practices, then, affect the original self and, as a result, such algorithmic (re)constitution strategies 'become constitutive of what they allegedly merely represent' (Pötzsch 2018: 3317). Nevertheless, because such constitution through governance is still distributed, there is (as yet) no totality or a master mould—instead, we are torn in multiple directions in accordance with the needs of the multiple choice architects and the data available to them. In other words, in the present posthuman context, humanness must be primarily seen not as some kind of metaphysical state of existence or a given presence but, first and foremost, 'as a process, a "becoming human" in connection with an environment and nonhuman actors' (Herbrechter 2013: 206), particularly non-human actors of a digital kind. But even that humanity is hybrid and fluid, as set by distributed algorithmic governance. The preceding also implies that agency should as well be seen as distributed and fragmented, bearing little resemblance to the singular autonomous acting self.

In addition, we are witnessing a new type of agency—'data-driven agency', the latter denoting 'a specific type of artificial intelligence, capable of perceiving environment and acting upon it, based on the processing of massive amounts of digital data' (Hildebrandt 2016: 4). Such new agents are not confined to a specific form or bodily structure (or lack thereof)—they could equally be physical robotic artefacts, such as autonomous vehicles, drones, smart robotic toys, companion robots, etc. and software agents, such as algorithms powering search engines or social networking platforms, increasingly intelligent bot profiles inhabiting such platforms, algorithmic generators of journalistic content, and many other types of actors or quasi-actors (Hildebrandt 2016: 4). Of course, their roles, functions, and capacities differ significantly but as all of them are becoming, or are projected to become, increasingly autonomous and learning from the environments that they are placed in, such code-enabled actors are bound to carve out a capacity to become independent causes on their own right.

There are also further ways in which human-centricity is a thing of the past, not least in the way in which human agglomerations, targeted advertisements, search results, or personalised recommendations simply cannot be adequately explained since the algorithms responsible for their provision are commercial secrets and, therefore, 'opaque, inaccessible, and unmodifiable' (Faraj et al. 2018: 63). Moreover, the 'network of networks' character of the internet underlies the permanently changing and shifting quality of the online environment in which infrastructure, software, and other elements can be layered, stacked, and repurposed, new game-changing elements can be added, etc., meaning that 'users must constantly renegotiate their positionalities, rights and abilities to interact with one another' (Iliadis 2018: 219–220). Indeed, a key difference is laid bare: in a traditional regulatory environment we typically know what the laws are because only the openly promulgated ones are in force. As a result, we can make our choices accordingly, in full knowledge and consideration. Code-as-law, meanwhile, is implicit, allowing us to, at best, register how *we are made to* act. But even the latter should not be taken for granted: in order for us to be conscious of being made to act, we also need to be aware of the alternatives and the ways in which such alternatives have been rendered unavailable. However, the narrow algorithmic if-then pathways provide a tunnel vision of what is available without the possibility of a broader picture.

Our increasing embrace and reliance upon intelligent code has fundamental implications upon our relationship with everyday artefacts. Indeed, whereas '[t]raditionally, technical extensions were used as a mere extension of human creativity and invention' and also 'as tools that help humans express intention', it becomes ever more evident that the present interaction with algorithms surpasses such one-way relationship, replacing it with a more egalitarian one, in which invention and creativity (and, arguably, at least to an extent, invention as well) are present on both sides (Zatarain 2017: 91). The structuring activity of algorithms becomes evident not only in relation to the affordances of the digital environment but also in terms of structuring of everyday life, its actions and routines. Digital calendars, for example, could be seen as an early illustration of such life-structuring code. More and more often, such calendars transgress the role of being mere mnemonic tools, effectively becoming personal assistants 'whose tracing capacities and behavioural algorithms can help users align their daily rhythms with ideals about efficient time management', nudging them to make prescribed lifestyle and behavioural choices (Wajcman 2018: 18). Similarly, the scheduling software that drives workplace planning can be seen as bringing the just-in-time economy to human relations, with the important caveat that 'instead of lawn mower blades or cell phone screens showing up right on cue, it's people' whose lives are made dependent upon 'the dictates of a mathematical model' (O'Neil 2016), thereby extending governance from online to offline. The issue of agency is glaringly open here: one has limited control of their schedule and, even more fundamentally, the rhythm and content of life, ceding that control to the calendar's algorithm or scheduling software, becoming subjected to whatever such intelligent code 'knows' its user needs.

Two seemingly contradictory processes are taking place at the same time: on the one hand, the algorithmic environment 'seeks to submit attention flows to needs and desires that will maximize financial returns' (Citton 2017: 73) while on the other hand, through constant bombardment with stimuli it 'inhibits our ability to be attentive towards others and attentive to our own desires' (Citton 2017: 142). That, nevertheless, is a contradiction only at first sight: True, the very purpose of data harvesting is being able to serve consumer desires, thus obtaining their affection and attention; but simultaneously such knowledge allows one to nudge humans and their desires towards predefined points, so that the object of desire and the way in which that object is desired are as close to the nudger's returns maximisation strategy as possible. Of course, that does not imply that data collection and analysis can cease at this point: one needs to still constantly observe the audience in order to keep track of the fluctuations of desire and to be able to nudge again if necessary.

5.2 Towards Posthuman Law

Due to algorithmic construction of choice architecture (and of the broader surrounding digital architecture), seemingly autonomous individual choice becomes highly debatable as 'people are given what they are likely to like, even if they have not chosen that', simultaneously making life more convenient and shutting individuals within predefined boundaries (Sunstein 2018: 257). And as individuals are effectively rec-reated as their own digital effigies on which any offering can be tested (Murdock 2017: 131), the offerings can be exquisitely moulded in order to have the maximum effect on one's view of the world (Boler and Davis 2018: 83) so that a person *cannot fail to choose*. Of course, the ques-tion of agency here rests on the already familiar dilemma: whether peo-ple are given the choice that they would have made out of their own rational self-interest, even without (or at least prior to) having realised that self-interest, i.e. whether such algorithms are allowing humans to be more Econ-like, or whether the decision is suboptimal for the person (but optimal for the choice architect instead) and the option was merely strategically placed to be chosen. If the former is the case, then it is still possible to claim that agency is preserved or even enhanced as people make *objectively better* choices and, thus, make better use of their agency (see e.g. Schmidt 2017: 405), while in the latter case agency is fictional only.

On the other hand, even if the choices are objectively better, algo-rithmic nudging, as opposed to that by democratically accountable gov-ernments, is susceptible to the objection from alien control: the power of some to affect the decisions of others puts those others under alien control, thereby reducing agency. And while in case of democratic gov-ernance the people at the receiving end of nudging at least have the opportunity to elect their nudgers, and so any nudging is, effectively, their own will (Schmidt 2017: 415), such democracy is obviously miss-ing from algorithmic governance. Under algorithmic governance, we are under the control of the algorithmic alien, for better or for worse, and there is nobody to hold the nudger accountable due to the typically opaque nature of corporate algorithmic governance.

There is, perhaps, also the question of practicality and efficiency: After all, 'because the intelligent software learns to know us better than we do ourselves, it is logical to cede control' (Wajcman 2018: 18), even if that significantly reduces our own agency. If an algorithm indeed delivers on what is expected, structuring our actions and nudging us towards choices that, in line with expectations, help maximise our performance and/or satisfaction, then ceding agency to intelligent code might easily become a self-reinforcing strategy. As we learn from both choice reinforcement, i.e. 'a tendency to make similar choices as in the past' as well as rein-forcement learning, which involves 'updating beliefs about the quality of default and non-default options based on past experience' (de Haan and

Linde 2018: 1204), if we routinely opt for something and our experience tells us it is good, then habit formation is the most likely outcome. In a way, one could perhaps even assert that the advent of algorithmic governance is likely to positively contribute to human agency by freeing up time and effort. Under this argument, having relieved us of the necessity to perform mundane time- and effort-consuming tasks as well as taken over the responsibility for routine choices, intelligent code will enable us to engage in happiness, community involvement, transcendental pursuits, and other noble activities, perhaps in a similar way in which slaves enabled Athenian democracy and philosophy (Etzioni 2017). In this, the reference to Hegel's master–slave dialectic, espoused at the beginning of this chapter, acquires new relevance in further underscoring the relationship between humans and algorithms.

Here it must be stressed that, despite all the schisms and divisions within the Western philosophical tradition, anthropocentrism has held a relatively central role, postulating human supremacy over the non-human, and resting on three major premises: that 'humans are special and privileged entities compared to other living beings (ontology), they are the only sources of knowledge (epistemology) and the sole holders of moral value (ethics)' (Ferrante and Sartori 2016: 176). Indeed, most of the thinking, at least since the Greeks, has been based on the 'supervening opposition between *physis* and *nomos, physis* and *techné*' (Derrida 1976: 33). As a result, *techné*, technology and machines in particular, were rendered only intelligible 'alongside or in opposition to the Human; as instruments, instrumental knowledges or as means-to-ends' (Seltin 2009: 48). In effect, though, as Mackenzie (2002: 43) shows, such a dichotomy tells us very little about the non-human counterpart but merely serves to stabilise an artificially rigid definition of the human. However, a different relation to technology is possible, one particularly pertinent to the assemblages of humans and code described above: of tools as technological exteriorisations of the human, thereby rendering the tool not the Other but an expression of the human (Stiegler 1998). Nevertheless, one should arguably take the argument somewhat further, stressing their mutual co-constitution: While humans make tools, expressing themselves through the latter, tools also make humans, i.e. the content of the human is affected by the subsequent use of tools.

It is, therefore, the goal of posthumanism to overcome deep-rooted anthropocentrism by postulating non-hierarchical interrelationship within nature (Ferrante and Sartori 2016: 177) as well as with the technological environment. Human mastery (or claim to it) is intentionally undermined, treating the material world not merely as something that passively reacts to the uniquely active humans; instead, it is implied that both the preconditions and the consequences of even the most outwardly conscious and rational actions are largely unforeseeable and unpredictable (Choat 2018: 1030). Current developments in the digital environment also enable (in fact, even demand) extension of such a line of thinking from the natural to the digital realm, locating in it a particularly potent form of co-constitution, which can already be intuited from the above.

Braidotti (2013) offers a crucial observation pertaining to the fluid nature of the concept of the 'human', rendering it contingent upon historical, cultural, and technological conditions. Not only we ourselves are likely to not always have been humans but also '[s]ome of us are not even considered fully human now, let alone at previous moments of Western social, political and scientific history' (Braidotti 2013: 1). To the first point, one only needs to think of, e.g. abortion debates and attempts to come to a conclusion as to at what point human life becomes protected while to the second, women, non-whites, non-heterosexuals, people of non-binary gender identity, etc. immediately spring to mind. In this sense, instead of being just one more fancy 'post-something' that typically retain the human at their heart, posthumanism is qualitatively different by putting into question 'the basic unit of common reference for our species, our polity and our relationship to the other inhabitants of this planet' and questioning the structures upholding a shared human identity in order to ultimately displace the hierarchy of species and 'a single, common standard for "Man" as the measure of all things' (Braidotti 2013: 2, 67). The world thus conceived clearly becomes more egalitarian and fluid.

A posthumanist view asserts that 'governing the world on the basis of the politics of modernity (top-down, cause-and-effect understandings) is dangerous, false and hubristic' (Chandler 2015: 850). Instead, an interpretation that embraces complexity and non-linearity must be offered. In a general sense, at the heart of posthumanism is the rejection of dualisms, such as that between humans and nature, opting instead for a worldview of humans as inseparable from their environment, including from other forms of life (Margulies and Bersaglio 2018: 103) and, increasingly, from algorithms and autonomous devices. As already seen in the previous chapters, the latter can be clearly demonstrated to take away a significant proportion of human agency by intervening in the causal chains of (primarily but not exclusively) social events in ways previously reserved solely to humans.

Notably, the traditional framework of human exceptionality can also be seen as ethically and morally problematic as it uncritically assumes a particular idea of 'human nature' the traits of which are, in fact, a 'concoction' of an exclusively European origin (Szollosy 2017: 156) whereas 'other religious/psychological traditions mobilize entirely different ways of thinking about these things' (Gunkel 2018: 75), postulating a much more closely knit relationship between the entirety of nature. As a result, traditional humanist anthropocentrism 'is not just insensitive to others but risks a kind of cultural and intellectual imperialism' (Gunkel 2018: 77). Moreover, even on occasions where human nature *is* presumed, that nature is being rendered 'plastic', malleable, and continually expanded by science, including through augmentation of the human species, thereby questioning what actually counts as human (Fukuyama 2002: 217–218). As a result, it is not at all surprising that at least some iterations of posthumanism, particularly the more environmentally minded ones, can be read as attempts to reassume this ethical responsibility (see e.g. Massumi 2014). However, the technological posthumanist strand appears to be of a more reactive nature, dealing primarily not with ethical imperatives (Gunkel [2018] being a notable exception) but, instead, with attempts to understand contemporary developments and (occasionally) propose appropriate responses.

Traditionally, the imaginary of a 'good' human life has been one freed from environmental risks, contingencies, and vulnerabilities— humans being seen not as part of nature and the animal world but as masters of them (Srinivasan and Kasturirangan 2016: 126). In other words, the quest for an 'insulated and protected life' as well as enhancement of pleasure and comfort through consumption at the expense of nature became the essence of human-centric development, leading to human exceptionalism, i.e. belief that 'as beings who have unique and exceptional qualities, humans deserve a standard of care that exceeds that of other beings', thus making instrumental use of others acceptable (Srinivasan and Kasturirangan 2016: 126–127). The current digital environment directly challenges such assumptions by rendering humans inextricably related to (and in many ways directly dependent upon) algorithms and other types of code.

In its algorithmic context, posthumanism should be seen in terms of technologically enabled existence or existence *within* technology. Such an acknowledgement does not necessarily imply surrendering oneself to some form of 'apocalyptic mysticism' (Herbrechter 2013: 3): Instead, posthumanism is better understood as a question, interrogating us whether we indeed know what being human is, means, and entails (Herbrechter 2013: 38) and simultaneously implying posthumanity to contain 'the aspects of network, complexity and emergence' (Herbrechter 2013: 206). The technological, and particularly digital, strand of posthumanism 'takes its basic unit of analysis as humans + tools, where tools would obviously include all forms of technology' (Mahon 2018: 25). Digital artefacts endowed with artificial intelligence or, at least, their own learning and decision-making capacity are 'explicitly posthuman' due to the erasure the human–machine distinction as they emulate (while also having roots in) human and animal biology (Mahon 2018: 80–81). Similarly, the body-data link, constructed through wearable devices of all kinds and the algorithmic analysis of such body data leads to 'a brutal regime of efficiency' (Greenfield 2018: 35), subjecting not only the mind but also the biological functions of the body to constant supervision and augmentation, once again erasing the distinction between natural and non-natural.

Moreover, of further relevance is the very outlook towards human beings that is illustrative of the overall spirit of algorithmic governance. Already in classical formulations of nudge, a key premise is that humans 'make predictable errors' and because such errors can be anticipated, 'we can devise policies that that will reduce the error rate' in order to make people better off (Thaler 2015: 325–326). Even if one accepts the benevolent Econ choice architect version (as opposed to the Iecons and the Becons), the nudge strategy thereby conceived serves as a licence for further datafication in order to determine choice patterns and 'true' objective interests in order to render the choice environment in such a way that the likely errors would still lead to the necessary outcome. In other words, it is data harvesting that produces the subjects (or digital effigies thereof) of code-as-law over the course of corporate and state surveillance, self-surveillance and tracking, and crowdsourced surveillance, even though such subjects are unstable and fluid, shifting every time new data become available (Lally 2017: 72; see also Cheney-Lippold 2017).

As evident from the above, the individual has no other option than playing second fiddle to predetermined outcomes and choice pathways. If in traditional nudging contexts the ones at the helm could still plausibly deny knowing (or at least expecting to know) 'what is best for everyone' (Thaler 2015: 326), such knowledge is now the essence of algorithmic governance. Nevertheless, such knowledge, even algorithmically derived, should not be seen as objective and value-neutral: as already asserted in this book, code-writing choices are never impartial, and this partiality is not to be overcome by simply expanding data pools or technological affordances. As Lally (2017: 72) puts it, 'while additional data and images or better algorithms might change how we understand a story, they are always mediated through and built upon particular epistemological frameworks'. Hence, 'best for everyone' is always still going to work out as what is best for those writing the code (or those on whose behalf the code is written).

As digital media condition the architecture and other environmental contexts of both social and psychological life, it might be possible to claim a certain unity between humans and such architecture, inferring as a result that 'emotions are becoming *with* social media rather than being controlled and dominated *by* them' (Tucker 2018: 39). Nevertheless, that would be true only in case the architecture was in an organic symbiosis with humans, unconditioned by purpose-written code and without a specific design for harvesting data and using the latter for behaviour-affecting purposes. Under the currently prevailing conditions, the relationship is much less egalitarian. Instead, as Danaher (2017: 72) suggests, 'developments in automation technologies may narrow the domain for genuinely meaningful action' whereby we may indeed live in an age of machine-enabled pleasure and abundance but be, nevertheless, merely passive recipients of the benefits generated by our machines without actively contributing to them. Hence, it is not only the idea of leisure, traditionally understood as simply an antithesis or absence of work, that is going to lose its meaning in the context of automation and will need to be reconsidered (Snape et al. 2017): the understanding of meaningful activities as such will also have to change along with the basic tenets of action.

Tellingly, it is most probably those that will be left behind, i.e. those who will not have become sufficiently posthuman, that are going to feel the shift in the most distressing way. Particularly that is going to be the case 'when the information necessary to comprehend and operate an environment is not immanent to that environment, but has become decoupled from it', becoming part of an augmented overlay instead, with such information likely including 'signs, directions, notifications, alerts and all other instructions necessary to the fullest use of the city', thereby relegating the unaugmented human to a second-class status (Greenfield 2018: 81). Hence, to reiterate from the first chapter, there is very little, if any, outside of datafication processes, and there will be even less of it in the future, leaving humans immersed in and transformed by algorithmic tools.

Technological posthumanism and the necessity for humans to share their agency with algorithms unavoidably raises the issue of the possibility and potential ways of conferring agency and, perhaps, even certain forms of personality to code-enabled actors. Such considerations are, however, beyond the scope of this book. Whether digital (bodyless or disembodied) actors are given personality on the same or similar basis as natural (see e.g. Gunkel 2018) or legal (see e.g. Papakonstantinou and De Hert 2018) persons, something in-between (having partial rights and obligations due to their autonomy but still tightly restricted by law due to ultimately being creations and agents of natural persons—see e.g. Čerka et al. [2017: 697]), entities progressing from simpler rights and obligations (such as insurance) to more complex arrangements, such as property rights (Tegmark 2017: 109), or no distinct personality at all, their influence is undeniable nevertheless.

At least in some cases, our attachment, manifesting itself as 'a multifaceted, meaningful relationship' of 'digital companionship' also plays a notable role (Carolus et al. 2018: 21). From the above, it can clearly be intuited that algorithms are at the core of today's life. True, their constitutive code still has to be written by humans but once such algorithms are put in place, they become the paramount nodes within the causal networks of human action and key elements within such networks in their own right. Moreover, by analysing data and (in)forming the lifeworlds and perceived interests of both the code-writers and their consumers, algorithms as a totality can be seen to have a significant causal input into the production of new algorithms, i.e. into their own (re)production as a digital species.

References

Ames, M. G. (2018). Deconstructing the Algorithmic Sublime. *Big Data & Society*. https://doi.org/10.1177/2053951718779194.

Beer, D. (2017). Envisioning the Power of Data Analytics. *Information, Communication & Society, 21*(3), 465–479.

Beer, D. (2019). *The Data Gaze: Capitalism, Power and Perception.* Los Angeles and London: Sage.

Boler, M., & Davis, E. (2018). The Affective Politics of the 'Post-Truth' Era: Feeling Rules and Networked Subjectivity. *Emotion, Space & Society, 27,* 75–85.

Braidotti, R. (2013). *The Posthuman.* Cambridge and Malden: Polity Press.

Bucher, T. (2017). The Algorithmic Imaginary: Exploring the Ordinary Effects of Facebook Algorithms. *Information, Communication & Society, 20*(1), 30–44.

Bucher, T. (2018). *If… Then: Algorithmic Power and Politics.* Oxford and New York: Oxford University Press.

Carlson, M. (2018). Automating Judgment? Algorithmic Judgment, News Knowledge, and Journalistic Professionalism. *New Media and Society, 20*(5), 1755–1772.

Carolus, A., et al. (2018). Smartphones as Digital Companions: Characterizing the Relationship Between Users and Their Phones. *New Media & Society.* https://doi.org/10.1177/1461444818817074.

Čerka, P., Grigienė, J., & Sirbikytė, G. (2017). Is It Possible to Grant Legal Personality to Artificial Intelligence Software Systems? *Computer Law & Security Review, 33,* 685–699.

Chandler, D. (2015). A World Without Causation: Big Data and the Coming of Age of Posthumanism. *Millennium: Journal of International Studies, 43*(3), 833–851.

Cheney-Lippold, J. (2017). *We Are Data: Algorithms and the Making of Our Digital Selves.* New York: New York University Press.

Choat, S. (2018). Science, Agency and Ontology: A Historical-Materialist Response to New Materialism. *Political Studies, 66*(4), 1027–1042.

Citton, Y. (2017). *The Ecology of Attention.* Cambridge and Malden: Polity Press.

Cotter, K. (2018). Playing the Visibility Game: How Digital Influencers and Algorithms Negotiate Influence on Instagram. *New Media & Society.* https://doi.org/10.1177/1461444818815684.

Danaher, J. (2017). Building a Post-Work Utopia: Technological Unemployment, Life Extension and the Future of Human Flourishing. In K. LaGrandeur & J. J. Huges (Eds.), *Surviving the Machine Age: Intelligent Technology and the Transformation of Human Work* (pp. 63–82). London: Palgrave Macmillan.

de Haan, T., & Linde, J. (2018). 'Good Nudge Lullaby': Choice Architecture and Default Bias Reinforcement. *The Economic Journal, 128,* 1180–1206.

Derrida, J. (1976). *Of Grammatology.* Baltimore: Johns Hopkins University Press.

86   I. KALPOKAS

Etzioni, A. (2017). Job Collapse on the Road to New Athens. *Challenge, 60*(4), 327–346.
Faraj, S., Pachidi, S., & Sayegh, K. (2018). Working and Organizing in the Age of the Learning Algorithm. *Information and Organization, 28*, 62–70.
Ferrante, A., & Sartori, D. (2016). From Anthropocentrism to Post-Humanism in the Educational Debate. *Relations, 4*(2), 175–194.
Fukuyama, F. (2002). *Our Posthuman Future: Consequences of the Biotechnology Revolution.* New York: Farrar, Strauss and Giroux.
Greenfield, A. (2018). *Radical Technologies: Thee Design of Everyday Life.* London and New York: Verso.
Gunkel, D. J. (2018). *Robot Rights.* Cambridge, MA and London: The MIT Press.
Hegel, G. W. F. (2018). *The Phenomenology of Spirit.* Cambridge and New York: Cambridge University Press.
Helbing, D., et al. (2017, February 25). Will Democracy Survive Big Data and Artificial Intelligence? *Scientific American.* Available at https://www.scientificamerican.com/article/will-democracy-survive-big-data-and-artificial-intelligence/.
Herbrechter, S. (2013). *Posthumanism: A Critical Analysis.* London and New York: Bloomsbury Academic.
Hildebrandt, M. (2016). Law *as* Information in the Era of Data-Driven Agency. *The Modern Law Review, 79*(1), 1–30.
Hornborg, A. (2017). Artifacts Have Consequences, Not Agency: Toward a Critical Theory of Global Environmental History. *European Journal of Social Theory, 20*(1), 95–110.
Iliadis, A. (2018). Algorithms, Ontology, and Social Progress. *Global Media and Communication, 14*(2), 219–230.
Janssen, M., & Kuk, G. (2016). The Challenges and Limits of Big Data Algorithms in Technocratic Governance. *Government Information Quarterly, 33*, 371–377.
Just, N., & Latzer, M. (2017). Governance by Algorithms: Reality Construction by Algorithmic Selection in the Internet. *Media, Culture and Society, 39*(2), 238–258.
Kelleher, J. D., & Tierney, B. (2018). *Data Science.* Cambridge, MA and London: The MIT Press.
Klinger, U., & Swensson, J. (2018). The End of Media Logics? On Algorithms and Agency. *New Media & Society.* https://doi.org/10.1177/1461444818779750.
Klowait, N. O. (2019). Interactionism in the Age of Ubiquitous Telecommunication. *Information, Communication & Society, 22*, 605–621.
Lally, N. (2017). Crowdsourced Surveillance and Networked Data. *Security Dialogue, 48*(1), 63–77.

Langlois, G., & Elmer, G. (2018). Impersonal Subjectivation from Platforms to Infrastructures. *Media, Culture and Society.* https://doi.org/10.1177/0163443718818374.

Lessig, L. (2006). *Code: Version 2.0.* New York: Basic Books.

Lupton, D. (2018). How Do Data Come to Matter? Living and Becoming with Personal Data. *Big Data & Society.* https://doi.org/10.1177/2053951718820549.

Lyon, D. (2014). Surveillance, Snowden, and Big Data: Capacities, Consequences, Critique. *Big Data & Society.* https://doi.org/10.1177/2053951714541861.

Mackenzie, A. (2002). *Transductions: Bodies and Machines at Speed.* London and New York: Continuum.

Mahon, P. (2018). *Posthumanism: A Guide for the Perplexed.* London and New York: Bloomsbury.

Margulies, J. D., & Bersaglio, B. (2018). Furthering Post-Human Political Ecologies. *Geoforum, 94,* 103–106.

Massumi, B. (2014). *What Animals Teach Us About Politics.* Durham, NC: Duke University Press.

Mayer-Schönberger, V. (2008). Demystifying Lessig. *Wisconsin Law Review, 4,* 713–746.

Mayer-Schönberger, V., & Cukier, K. (2017). *Big Data: Th Essential Guide to Work, Life and Learning in the Age of Insight.* London: John Murray.

Murdock, G. (2017). Mediatisation and the Transformation of Capitalism: The Elephant in the Room. *Javnost—The Public, 24*(2), 119–135.

O'Neil, C. (2016). *Weapons of Math Destruction.* New York: Crown.

Papakonstantinou, V., & De Hert, P. (2018). Structuring Modern Life Running on Software: Recognizing (Some) Computer Programs as New '*Digital Persons*'. *Computer Law and Security Review, 34*(4), 732–738.

Pötzsch, H. (2018). Archives and Identity in the Context of Social Media and Algorithmic Analytics: Towards an Understanding of iArchive and Predictive Retention. *New Media & Society, 20*(9), 3304–3322.

Reagle, J. M. (2019). *Hacking Life: Systematized Living and Its Discontents.* Cambridge, MA and London: The MIT Press.

Schmidt, A. T. (2017). The Power to Nudge. *American Political Science Review, 111*(2), 404–417.

Schraube, E. (2009). Technology as Materialized Action and Its Ambivalences. *Theory & Psychology, 19*(2), 296–312.

Schwitzgebel, E., & Garza, M. (2015). A Defense of the Rights of Artificial Intelligences. *Midwest Studies in Philosophy, 39*(1), 89–119.

Seltin, J. (2009). Production of the Post-Human: Political Economies of Bodies and Technology. *Parrhesia, 8,* 43–59.

Snape, R., et al. (2017). Leisure in a Post-Work Society. *World Leisure Journal, 59*(3), 184–494.

Srinivasan, K., & Kasturirangan, R. (2016). Political Ecology, Development and Human Exceptionalism. *Geoforum, 75,* 125–128.

Stiegler, B. (1998). *Technics and Time, 1: The Fault of Epimetheus.* Stanford: Stanford University Press.

Sunstein, C. R. (2018). *#Republic.* Princeton and Oxford: Princeton University Press.

Szollosy, M. (2017). ESPR Principles of Robotics: Defending an Obsolete Human(ism)? *Connection Science, 29*(2), 150–159.

Tegmark, M. (2017). *Life 3.0: Being Human in the Age of Artificial Intelligence.* London and New York: Penguin Books.

Thaler, R. H. (2015). *Misbehaving: The Making of Behavioural Economics.* London and New York: Penguin Books.

Thomas, S. L., Nafus, D., & Sherman, J. (2018). Algorithms as Fetish: Faith and Possibility in Algorithmic Work. *Big Data & Society.* https://doi.org/10.1177/2053951717751552.

Tucker, I. (2018). Digitally Mediated Emotion: Simondon, Affectivity and Individuation. In T. D. Sampson, S. Maddison, & D. Ellis (Eds.), *Affect and Social Media: Emotion, Mediation, Anxiety and Contagion* (pp. 35–41). London and Lanham: Rowman & Littlefield.

Vaidhyanathan, S. (2018). *Anti-Social Media: How Facebook Disconnects Us and Undermines Democracy.* Oxford and New York: Oxford University Press.

Vladeck, D. C. (2014). Machines Without Principals: Liability Rules and Artificial Intelligence. *Washington Law Review, 89*(1), 117–150.

Wajcman, J. (2018). The Digital Architecture of Time Management. *Science, Technology and Human Values.* https://doi.org/10.1177/0162243918795041.

Zatarain, J. M. N. (2017). The Role of Automated Technology in the Creation of Copyright Works: The Challenges of Artificial Intelligence. *International Review of Law, Computers & Technology, 31*(1), 91–104.

CHAPTER 6

Dis-imagining Rights, Legitimacy, and the Foundations of Politics

Abstract In this chapter, the recent changes in governance are considered from rights- and legitimacy-based perspectives. While not aiming at an exhaustive treatment of the matter, this chapter teases out some emergent issues that merit further consideration. Although, if a post-humanist framework is accepted, it is no longer reasonable to rely on a *human* rights-based assessment, it is important to show how some otherwise taken-for-granted assumptions are rendered questionable today. For that reason, some provisions from the Universal Declaration of Human Rights (UDHR) and the International Covenant on Civil and Political Rights (ICCPR) are compared against the current trends in algorithmic governance. Meanwhile, a consideration of legitimacy, particularly in its political sense, helps uncover the emergent fault lines within the three-way interplay between the electorates, the public authorities, and code.

Keywords Human rights · Political rights · Legitimacy · Agency · Hybrid governance · Dis-imagined communities

As is clear from the preceding chapters, it has become largely impossible to escape algorithmically enabled processes and algorithmic agents broadly conceived—particularly as the cost of deploying them is decreasing steadily while the benefits remain high (Bodo et al. 2017: 136). Moreover, such uptake is also further motivated by progress in algorithmic technology and the increasing capacity to automate tasks that are

© The Author(s) 2019   89
I. Kalpokas, *Algorithmic Governance*,
https://doi.org/10.1007/978-3-030-31922-9_6

both difficult and nuanced, e.g. moving from textual search to image and voice recognition. As a result, algorithms, particularly of a machine learning kind, are becoming part of the critical infrastructure of our societies, simultaneously making any 'human in the loop' not only expedient but also counterproductive (Edwards and Veale 2018: 26). The latter opens up ample opportunities for automated algorithmic governance that has become progressively manifest in private regulation but is increasingly moving into the domain of public regulation as well. While some (see e.g. Mantelero 2018: 771) are adamant that a *human* rights-based assessment of algorithmic governance is necessary, the posthumanist outlook laid down in the preceding chapter should have already made the futility of such an endeavour clear. Instead, it is important to focus on the shifts within traditional rights- and legitimacy-based constraints on governance that should serve as an impetus for new and creative reconsiderations of novel ground rules for sociopolitical life.

6.1   Being Governed: Further Observations

Certainly, the striving for and adoption of 'data-driven decisions' in both the private and, progressively, the public sectors has been enabled not only by the combination of large datasets, cheap computation, and advances in data-crunching capacity but also by political goals, particularly as an attempt to automate and ostensibly augment decision-making processes in areas as diverse as 'justice, policing, taxation or food safety' (Edwards and Veale 2018: 26–27) where the alleged efficiency and value-neutrality of code become crucial selling points, helping to win public support. Nevertheless, at the same time an apt question remains that of attribution of agency and responsibility in situations when the digital agents involved 'are "obfuscated by nature", or when algorithms function as cultural techniques that seem to have their own agency' (Markham et al. 2018: 5). As a result, this non-human side of social, economic, political, legal, etc. life merits special attention, however transgressive that might be from the standpoint of currently established perspectives that may not be so easily compatible with the state of being 'co-constituted by and with the non-human' and the potentially radical implications that such a shift involves (Cudworth and Hobden 2013: 655, 659). In particular, situations (increasingly commonplace) when algorithms, deployed as (relatively) autonomous actors, effectively come to possess their own agency and the capacity to make quasi-moral choices (Just and Latzer 2017: 255), cannot but provoke questions of accountability and, much more fundamentally, the place of humans in legal and political matters.

Overall, as shown in the previous chapter, the status of humans as exclusively active agents must be questioned, focussing instead on embeddedness in human-digital assemblages. However, there are further issues, stemming partly from the fact that data, including of various sensitive kinds, are left 'in the hands of controllers who may be difficult to identify' (Edwards and Veale 2018: 34–35) thus, among other things, making it almost impossible to keep track of and foresee potential influences and regulatory constraints, some of which are deeply implicit. At the same time, algorithmic agents ever more often turn from mere arbiters to active agents shaping our interactions both with and in the wider environment (Bodo et al. 2017: 136). Simultaneously, it would be futile to conceptualise a separate 'datasphere' (as in e.g. Bergé et al. 2018) because the physical and the digital are simply inextricably intertwined. Instead, it is the broader posthumanist framework of 'co-constitution' of multiple actors and conditions and irreducible multiplicity of causality (see e.g. Mitchell 2014: 7) that counts.

The very fact that even the seemingly minuscule choices are guided, and often even determined, either through digital-architectural enabling/disabling or through the commonplace and seemingly mundane feature or recommendation that ultimately entangle us within a web of pregiven values and judgement (Tufekci 2019) demonstrates that we are, to a significant extent, lost within the broader posthuman assemblages. No less importantly, such technical rules—or, more precisely, the data generation and analysis processes enabled by such rules—enable the curation of content pertaining to anything from well-being and fitness to news and entertainment, which is, in turn, 'fed back to us in a self-affirming loop of our habits', thereby providing for 'behavioural maintenance or change' (Iveson and Maalsen 2019: 333). That, in turn, brings back the matter of the capacity to decide. Simultaneously, however, the agendas of such algorithmic agents are 'unknowable and unchangeable' (Scott 2019), at least for actors outside (the largely self-contained) technology industry. That is in stark contrast with more traditional modes of regulation, whence at least lip service to a (more or less) explicitly formulated public interest is key to acceptance of measures as legitimate. Moreover, whereas normally laws and other regulations must be knowable in advance, in this case they become either opaque or constantly shifting, revealing themselves only within the process of humans going about their daily business.

A further point to note is that the emergent, posthuman, modes of control are not tied to individuals and are not geared towards the control *of* individuals as either social beings (see e.g. Foucault 2012) or as biopolitical dominance over bare life (see e.g. Agamben 1998). Instead, 'newly emerging modes of social control [...] operate through the *modulation of datafied dividuals*', whereby 'each database offers the opportunity to control the social via modulatory means, acting not on individuals but via the distribution of personhood into dividuals monitored and modulated by different digital systems' (Iveson and Maalsen 2019: 334). In a characteristically digital sense, the inviolability of the person is put into question here. Certainly, it is not the bodily integrity that matters here but the conception of the person as a coherent entity in relation to the law (or other forms of governance) in light of the new capacities of disassembling and partially reassembling as necessary. In this sense, once one begins approaching algorithmic law as a distinct phenomenon, the paramount principle of equality of persons before the law (see, prominently, UDHR Article 7) becomes shaky. In particular, that relates to differences in availability of data input on subjects and the ensuing differences in attributes and variables in accordance to which subjects can be dis- and reassembled, resulting in divergent degrees of regulability.

One more thing to remember is the automatic creation of data that elides the user's wilful decision and 'is structured in such a way as to extract, permanently, the greatest possible amount of data from each access' (Bergé et al. 2018: 152). To an unprecedented extent, data can even be seen as self-generating, whereby first-order data give rise to metadata (e.g. location of generation and other environmental and contextual characteristics), thus engulfing every action in datafication processes (Bergé et al. 2018: 153). In addition to the widely discussed privacy issues, there is also a notable shift in the origin of law- and rule-making. Crucially, due to the highly technical nature of the data collection and analysis infrastructure, one can identify as a characteristic feature of contemporary societies the condition of 'being governed by technical rules embedded in the activity or in the product' (Bergé et al. 2018: 155). In other words, it is the features embedded in the technological or digital design of the devices, platforms, and other products and artefacts that determine what kinds of variables and considerations (i.e. types of data) become the input of governance measures. Such questions become largely detached from both politics and policy (and, therefore, removed from scrutiny) and are, instead, handed over from bureaucrats and politicians to hardware and software engineers. As a result, the modes of data collection and, coextensively, input into regulation can easily be swayed by efficiency and profitability rather than rights (especially privacy) and legitimacy considerations.

Indeed, as Lessig stressed back in 2006, one must constantly probe the questions of '[h]ow the code regulates, who the code writers are, and who controls the code writers' in order to understand the basic features of contemporary life (Lessig 2006: 79). Definitely, that is not some kind of aberration but simply a matter of technological change. In fact, 'in the middle of the nineteenth century the threat to liberty was norms, and in the start of the twentieth century it was state power, and during much of the middle twentieth it was the market', all having exerted paramount structuring and organising power upon societies in their heyday; in the same vein, 'in the twenty-first century it is a different regulator – code – that should be our current concern' (Lessig 2006: 121). The matter becomes even more pressing when such regulation through code permeates both private and public settings. While private regulation through code is relatively straightforward—as in the architecture of platforms that enable and constrain actions online—the public incarnation might be slightly less intuitive. Nevertheless, at least two ways of code's intrusion into public regulation can be identified, both already intuited in the preceding chapters. One is the relatively indirect influence that opinion, attitude, and lifestyle formation as well as struggle for attention has on political processes and the making of law. The other is the direct employment of code for anything from predictive policing to generation of policy decisions to (potentially) creation of legal frameworks that adjust to individual subjects.

In the private sense, code-as-law has, as already demonstrated, an increasing role in regulating everyday life, structuring communication channels, determining information acquisition, etc. But no less importantly, machine learning algorithms become involved in regulating welfare and freedoms due to their use 'to arrive at decisions vital to individuals, in areas such as finance, housing, employment, education or justice', meaning that their vital role now spans 'private, public and domestic sectors of life' (Edwards and Veale 2018: 19). Hence, depending on one's attributes and personal history (and the personal histories of those deemed to be similar to that person), one would be ascribed to categories, groups, tiers of risk, or other constructs. Such ascription would, in turn, mean that certain opportunities would be allowed or disallowed, actions prohibited or made legal to specific individuals and not as a matter of blanket provision. Nevertheless, despite such importance of code, public control and supervision is lacking, first and foremost due to the opacity (and, particularly in case of machine learning algorithms, dynamic nature) of the algorithmic provisions in question (Chenou and Radu 2019: 80).

In relation to the above, one must keep in mind, in line with Article 10 of UDHR, that '[e]veryone is entitled in full equality to a fair and public hearing [...] in the determination of his rights and obligations' as well as charges. A very similar provision, particularly in relation to being promptly informed about the nature of the charges, is also enshrined in Article 9(2) and, even more forcefully, Article 14(3a) of ICCPR: According to the latter provision, everyone is entitled '[t]o be informed promptly and in detail in a language which he understands of the nature and cause of the charge against him'. However, the latter is significantly impeded due to the lack of access to and understanding of the code behind the regulation applied (elaborated in greater detail in the preceding discussion of algorithms as 'black boxes'). Certainly, one might counterargue that being charged with an offence and having regulation algorithmically moulded to one's person are completely different matters. Nevertheless, algorithmic limitation of possible actions and (social, political, or legal) opportunities can potentially be seen as even more serious than a charge—it is, in its nature, already a sanction, albeit not one handed by a competent judicial institution but by and in code (and, hence, potentially even more problematic).

In the above context, it is troubling that machine learning algorithms are typically 'complex and incomprehensible to humans', their creation being concerned 'with predictive performance rather than interpretability as a priority' (Edwards and Veale 2018: 26). In fact, the complexity is often such that even the developers themselves are finding it extremely difficult to understand the development (particularly in machine learning) and performance of their own algorithms (Riley 2019: 27). Hence, a much broader discussion on the necessity and feasibility of 'new norms, new rules, and new rights for the individuals' (Chenou and Radu 2019: 80) is necessary. On the other hand, once such rights against the wielders of data and code are created, public actors only further entangle themselves with private ones by passing 'strong regulations they may not be able to implement themselves without the collaboration of private actors', the result being that 'private actors are given new responsibilities in the governance of technologies and technologically-enabled markets' (Chenou and Radu 2019: 96–97). After all, it is becoming clear that the largest platforms, such as Google and Facebook, are themselves becoming hybrid, blending 'the features of commercial platforms and public infrastructures' (Kreiss and McGregor 2019: 3). Similarly, as Lambach (2019: 18) contends, digital territories have become corporate territories anyway, necessitating that corporations let states in, leading to co-governance, whereby 'corporate and state territories intersect, allowing states to delegate the enforcement of laws to corporate governors' (for a similar argument, see also Susskind 2018: 155). Perhaps most prominently, through the implementation of the 'right to be forgotten' in the European Union, 'Google became inserted in the European legal system as a first instance to look at the cases of online privacy protection triggered by individual requests' (Chenou and Radu 2019: 97). In this way, private regulation through code (what can and cannot be retrieved by a search algorithm) has become both a matter of public regulation and a means of inclusion of private actors *in* public regulation, pointing to new, hybrid, forms of governance that render the public and the private inseparable, both involved in the generation and application of increasingly interrelated algorithmic law.

A further problematic element can be traced to the dynamic nature of algorithmic law: as is implicit in underlying machine learning techniques, law-generating algorithms 'will gradually change and respond to new information to refine prediction accuracy' (Robinson 2017: 299). Indeed, it is becoming evident that we are witnessing, broadly speaking, 'a shift from law enforced by *humans* to law enforced by *digital systems*' through laws that are unbreakable due to being 'encoded in the world around us' and tailored to one's life circumstances and track record (Susskind 2018: 101, 105–107). While, in a traditional setting, law would be passed by a competent authority and would subsequently remain relatively stable (or a sanction, once final, would remain in place and unchanged), algorithmic law is characteristically dynamic. Also, whereas traditional laws can also be changed and updated in light of important new developments not yet captured in law (or when the nature of already captured elements changes significantly), such change is, in all likelihood, not a permanently ongoing process. Furthermore, any changes to existing legislation have to be promulgated prior to taking effect. Hence, in case of traditional law, one should (and even has an onus to) know what they are, and will be, allowed (or not allowed) to do. The same is not the case with algorithmic law that changes in real time. Hence, one is likely to encounter an asymmetry of power: while the algorithms are predictive with regards to human behaviour, the ability of *humans* to predict their environment and limits on action (i.e. the behaviour of algorithms) is severely reduced. Hence, the question of legitimacy of predictive algorithmic rule is left glaringly open.

The above quality of algorithmic law must be put into question by relating it to Article 11 of UDHR and, particularly, the presumption of innocence (also the feature of Article 14(2) of ICCPR) since, being anticipatory, predictive algorithms must operate on a presumption of guilt-soon-to-come. While the latter is not *strictly* identical to a presumption of guilt plain and square, it may still have the same detrimental effects, forcing the person to fight an uphill battle against not even the apparatus of state but an impersonal and opaque algorithmic system. After all, it seems natural to suggest that if an algorithm decides to disallow a course of action, then it must hold that the person in question (almost certainly) *will* commit a culpable act *if only given the chance*, so must be punished *in advance*. Likewise, as stipulated in Article 11(2) of UDHR, nobody should be accused of an act that was not prohibited at the time it was committed. Nevertheless, due to the malleable nature of algorithmic law that changes in real time, acts can often become prohibited *simultaneously with* the performance of that act. As a result, the person can no longer be sure whether their intended course of action is in line with norms and regulations and how that is going to affect their standing in the private (e.g. within a platform) or public pecking order. And as machine learning techniques can be employed to tie together even information that might otherwise appear to be unrelated, effectively any human action could positively or negatively contribute to their digital representation on which the opening or closing of doors depends (Robinson 2017: 295), potentially rendering such code-as-law a matter of control, as exemplified by China's social credit system.

Definitely, it is crucial to also keep in mind that algorithmic law is not *unique* in its tailored nature: traditional laws are tailored as well, applying to some groups, professions, ages, etc., meaning that the novelty is, instead, one of degree, with algorithmic tailoring focusing on smaller groups and, potentially, individuals, regulating them in finely calibrated ways (Robinson 2017: 298). Indeed, in the emergent circumstances, law comes to apply to the person in different ways not because of some predetermined characteristics (age, employment type, family circumstances, etc.) but as a result of one's behaviour and the behaviour of those deemed similar. Hence, once again one encounters the difficulty of knowing in advance what is and what is not allowed, what provisions one is and is not subjected to, except by trial and error, with errors leading to even further degradation of one's standing.

In a general sense, algorithmic law would perhaps enhance the freedom of those deemed to be harmless by allowing them to do more things than before or reducing the number of necessary checks while simultaneously curtailing the freedom of those deemed to be dangerous or, at least, unreliable by disallowing things that would normally be allowed (Robinson 2017: 317). Of course, one of the key matters here would be an issue already discussed earlier in this book: that of accuracy with which algorithms are capable of identifying those who must be sanctioned. But even the question of accuracy of algorithmic prediction notwithstanding, there is also a political threat as well since 'tailoring ensures that oppressive restrictions only apply to few individuals and will not affect many voters' and, no less importantly, 'the idea that an algorithm *determines* a labelled group to be *risky* [...] adds legitimacy to oppressive restrictions' (Robinson 2017: 318). Nevertheless, the imperative of Article 7 of UDHR, that '[a]ll are equal before the law and are entitled to equal protection against any discrimination' (see also ICCPR Articles 14 and 26) must be kept in mind. Particularly as algorithms typically operate as a matter of sorting, being ascribed to groups based on even the most obscure shared attributes and without any will or intention to associate with that group can easily be a matter of severely restricting or expanding one's status and capacities.

6.2   The Legitimacy of Algorithmic Governance

In addition to personalisation of regulation, we are also facing personalisation of experience, whereby 'an increasing fraction of our digital experience is unique to us, and unknown to others, insulating our digital experiences into individual experience cocoons', weaved by 'opaque algorithmic agents' (Bodo et al. 2017: 139). Because such experiences are simultaneously unique to the person (tailored to their digital profiles) and private (typically experienced individually, e.g. while swiping through one's social media feed), our personalised cocoons 'remain unknown – perhaps even unknowable – and thus incommensurable for everyone else', with family members and neighbours no longer sharing the same (or even similar) exposure to events and interpretations (Bodo et al. 2017: 140). Notably to this context, Article 19 of UDHR mandates that everyone possesses the 'freedom to hold opinions without interference and to seek, receive and impart information and ideas'. An analogous provision is also present in Article 19 of ICCPR. While at first sight the connection might be unclear, it is important to stress that opinion formation becomes constrained. While it is no longer government censorship that is the threat, it has been replaced by algorithmic ascription to preference silos, thereby wrapping individuals inside convenient opinion-congruent information environments. Again, it must be acknowledged that previously news editors would not shy away from tailoring their offering to particular agendas. Nevertheless, rarely (if ever) did they have the capacity to engulf their audiences so completely as do the code-architects of today's social-mediated environment. As a result, the freedom to seek and receive information and, hence, to hold opinions, exists only insofar as it is permitted by code.

It is also crucial to note that, inherently to the posthumanist argument, 'No longer are the key subjects of political power the individual and the mass. Rather, individuals have become "dividuals", not constituting a "mass" but rather distributed across samples and databanks' (Iveson and Maalsen 2019: 334). As can be intuited, the freedom of association is here rendered problematic in a way similar to the challenge to the freedom of opinion above. Most notably, freedom of association (Article 22 of ICCPR), not only in terms of non-interference with creating associations but also as in protection from forced association, comes under threat particularly in relation to the algorithmic creation of groups that may not even be evident to those ascribed to them as well as in the architecturally embedded inability (or, at the very least, severe hindrance) to associate with those algorithmically deemed dissimilar.

Closely interrelated with the above is also perhaps the most intuitive account of legitimacy—one focusing on 'the right to rule and the recognition by the ruled of that right' in order for institutions 'to develop, operate and reproduce themselves effectively' (Jackson et al. 2012: 1051) as well as 'to justify, in moral terms, the wielding of [...] power' (Coglianese 2007: 160) as 'appropriate, proper, and just' (Tyler 2006: 376). However, in present circumstances, it becomes unclear *to whom* such a justification has to be made. If it is difficult to conceive of a national public that would be a referent to the government—and, in the context of algorithmic personalisation of law and other applicable regulations—of the public sphere. Instead, we seem to be witnessing a process of dis-imagination of communities, to paraphrase Anderson's (2006) famous notion. If that is truly the case (as it seems to be), then a broad recognition of legitimacy might appear to no longer be possible (provided it ever was). Instead, it makes ever more sense to seek legitimacy within those algorithmically ascribed information-association silos without much regard to the broader community, as illustrated by the propensity of today's political agents to foster and subsequently exploit polarisation.

Clearly, it must by now be uncontroversial that key domains and characteristics of politics, including but not limited to 'political systems, elections, decision-making, and citizenship', have increasingly become 'driven by aspects or by-products of automation or algorithmic systems at different systematic levels' (Ünver 2019: 1). To an extent, the preceding can be framed as a matter of responsiveness, which can, perhaps, be significantly enhanced because 'political parties, governments, and leaders across the world employ some versions of text mining algorithms to keep up-to-date with citizen and voter sentiments and issues that they want to be raised', opening up the no longer unrealistic possibility for 'AI-based recognition systems to learn and designate critical issues much faster and more reliably, enabling real-time and fluid agenda-setting capability to politicians' (Ünver 2019: 9). Clearly, should we assume a rather restrictive view of legitimacy, simply as '[t]he substantive alignment of policy content with the dominant attitudes of society' (Wallner 2008: 437), then the legitimacy of political processes and institutions could be even dramatically enhanced. However, in addition to potential normative criticism of self-referential implosion of strong immanence (see Kalpokas 2018), there is also a practical problem of inclination towards solutions that are known to be popular instead of those that may be objectively necessary. Thirdly, not least, there is a broader question of the meaning of vote: Essentially, what becomes of a voting decision if the policies promised and adopted are known in advance to nudge individuals towards embracing a particular voting behaviour, potentially to the extent of the voter no longer being able to *fail to choose* (as demonstrated in Chapter 4).

Notably, the procedural level becomes important here. The latter is typically taken to encompass 'democratic accountability, with elections being the principal defining characteristic, and also in terms of institutional arrangements like separation of powers, transparency, and rule of law principles intended to combat abuses of power' (Coglianese 2007: 161). All of those can be formally upheld, even though rule of law, as shown above, might be conceived as rule of the algorithm through interpretation of data while separation of powers can be undermined by the same private authors of code being simultaneously embedded in multiple institutions as a matter of hybridisation of governance; likewise, the freedom and fairness of elections can be undermined by the enhanced capacities to nudge voters towards predetermined choices. In addition, the procedural level also must be supplemented by the substantive one, the latter being 'usually defined in terms of rights, typically rights enshrined within a constitution that makes certain actions off limits even to an otherwise procedurally legitimate legislature' (Coglianese 2007: 161). While traditionally it would have been more readily conceivable for a decision to be procedurally legitimate and substantially illegitimate (as in a democratically elected parliament passing an anti-constitutional law), with algorithms added into the mix it is not unlikely to get substantially legitimate decisions with procedurally dubious premises.

Moreover, there is a clear and notable shift in the public–private interaction when it comes to government and governance: As correctly observed by Chenou and Radu (2019: 80), '[e]ven if Big Data is also a public phenomenon, given the amount of data collected by public administrations, the capacity to make sense out of the data requires the intervention of private tech companies'. In other words, public institutions, including the government itself, become dependent on their private partners (and, essentially, counterparts in governance). Indeed, one can observe a significant redefinition of governance practices as such: for example, '[b]y making available information contained in various physical sites, by organizing and classifying it, private intermediaries [...] play a crucial role not only in big data analytics but also in governance processes more generally', ultimately leading to a condition whereby 'state law is nowadays efficient only if complemented by private orderings' (Chenou and Radu 2019: 80). The result is a 'hybridisation' of governance, ensuing from '[t]he specific characteristics of the digital era', such as 'monopolistic markets, private expertise, and private ownership of big data' (Chenou and Radu 2019: 96). Hence, one can witness significant non-independence of government institutions and an irruption of practices based on considerations other than the public interest into the public regulation and governance domain as well as quasi-panoptical visibility rendered possible by such hybridisation.

Certainly, should one assume that 'political legitimacy ought to be the ultimate goal of any system of governance' (Rothstein 2009: 326), i.e. that it is not the premise but the results actually achieved that give rise to legitimate government, then the fulfilment of this teleological criterion can perhaps be brought closer through public–private interaction in governance. The primary reason is, again, the increased capacity for responsiveness to the wishes, expectations, and desires of the publics (i.e. target fragments of the dis-imagined community) that are rendered knowable and measurable. Susskind (2018: 247–248) is onto something very similar with his idea of Data Democracy, whereby decision-making would be carried out 'on the basis of data rather than votes' as data collection and analysis would lead to the creation of 'the sharpest and fullest possible portrait of the common good', making it 'a *really* representative system – more representative than any other model of democracy in human history' (Susskind 2018: 248). Nevertheless, this account does nothing to address a fundamental problem stressed by, e.g. Zuboff (2015): the inequality between those who collect data and put them to use and those *whose* data are collected. Likewise, the more grassroots-focused perspectives on legitimacy are turned on their head. Should we assume, then, that legitimacy rests on the feeling of involvement and the ensuing worthiness of participation, congruence with personal beliefs about the actors involved, and perception of fairness on both procedural and distributive levels (see e.g. Weatherford 1992: 151), then the process of legitimation shifts from opinion formation based on government performance to government performance (or, at least, the presentation thereof) being moulded in accordance with datafied building blocks of opinion, potentially even prior to that opinion becoming formulated by the target audience. This can, among other things, be seen as one of the key premises of so-called post-truth or post-factual politics (see e.g. Kalpokas 2019).

Surely, there can also be efficiency benefits when it comes to formulating actual policies because 'AI can help in optimizing scenarios by running multiple tandem processes of outcome and resource distributions, as well as forecasting possible types of public reaction against these outcomes' (Ünver 2019: 10). A crucial problem in this context is that '[a]lgorithms blur the separation between the decision and the process by which that option prevailed over others', with the ultimate result that 'the policy crafting process, as well as the decision itself, becomes detached from political legitimacy and sovereignty considerably' (Ünver 2019: 10). In other words, refashioning of decision-making processes from algorithmically enhanced to algorithmically enabled displaces the typical loci of power. This shift, in turn, has a much broader effect on political practices and taken-for-granted paradigms, particularly problematising the location of authority for making binding decisions.

The above, in turn, leads to a pressing question for the future: 'Will the citizens vote for an algorithm, or a series of algorithms, or a human decision-making team presiding over algorithmic structures that make most of the critical decisions?' (Ünver 2019: 8). But even without such a dramatic shift, the hybridisation of government and governance implies that '*thousands upon thousands* of decisions [...] will be taken every day, decided automatically and executed seamlessly with no right of appeal' (Susskind 2018: 191). At the very least, the result of hybridisation and digitisation of governance, regulation, and enforcement—'[d]igital law, privatized force, autonomous systems of power'—are more than likely to profoundly shake the foundations of sociopolitical life (Susskind 2018: 121). Crucially, then, we may end up witnessing dual processes of campaigning being increasingly premised on audience data but actual decisions being increasingly alienated, made through public–private hybrid systems only one part of which would be accountable to the populace.

Again, the question of legitimacy looms large. On the one hand, it is not uncommon to assume that cognitive legitimacy is dependent upon 'the degree to which an organizational form is taken for granted', implying that the perception of an innovation must be premised upon the degree to which it is accommodated within pre-existing cultural schemata (Suddaby and Greenwood 2005: 37). In this case, the emergent forms of hybrid governance are unlikely to be well-perceived. Likewise, if treating legitimacy from a normative perspective 'as *the right to rule*' whereby 'a law-making institution possesses the right to rule if it is morally justified in *issuing laws and rules* and *attempting to make individuals comply with those rules* (usually through coercive means)' while 'individuals have *reasons to comply* with the law-making institution or (even more strongly) *a moral obligation to obey*' (Machin 2012: 102), automated digital processes do raise doubts, including concerning obeyance to algorithms. Nevertheless, such doubts are premised on the presence of choice, which cannot be taken for granted: in today's world, where the digital is increasingly interconnected with the physical and the virtual is in ascendance, it is becoming more and more difficult to escape algorithmic moulding of life, regardless of whether such moulding is private or hybrid.

On the other hand, one might counterargue by returning to Rothstein's (2009) argument that the procedural element should perhaps not be treated overly seriously. After all, it is suggested, 'electoral democracy is highly overrated when it comes to creating legitimacy' while, in actual practice, 'legitimacy is created, maintained, and destroyed not by the input but by the output side of the political system', effectively depending upon 'the quality of government, not the quality of elections or political representation' (Rothstein 2009: 313). In a somewhat related manner, d'Aspremont (2006: 880) distinguishes between *'legitimacy of origin* and the *legitimacy of exercise'*, treating both as completely separate, implying that political authorities can be independently legitimate in one way or the other (or, ideally, both). Also similarly, Muirhead and Rosenblum (2019: 33) attest to two senses of legitimacy, this time labelling them the philosophic and the sociological ones: whereas the former 'asks what kind of regime, in principle, would be worthy of support', the latter is concerned with 'whether citizens in fact view their political order as worthy of their support'. Nevertheless, relying on this distinction, whatever its iteration, is not without problems either.

Certainly, legitimacy-qua-public-support does not automatically imply the use of government power for aims beneficial to either the state itself or some wider network of shareholders (including humanity as such)—merely that power is accepted and *perceived* to have everybody's best interests in mind (Levi and Sacks 2009: 355–356). After all, politics entails the necessity of choosing on an undecidable terrain whereas, on the one hand, the successful side 'must have reflected an otherwise neglected aspect of the political community's existence' but, simultaneously, such reflection is only an ex post facto attribute of success (Kalpokas 2018: 158). Similarly, as Honig (2009: 47) claims, 'victorious political actors *created* post hoc the clarity we now credit with having spurred them on to victory ex ante'. In other words, certainty is created as if there was a firm anchoring point (and, therefore, a 'naturally' desirable outcome) within the perpetual mutual reproduction of the constitutive and constituted strands of the political process (Kalpokas 2018: 162, 179).

What often happens, though, is the 'naturalisation' of an outcome that happens to be desired by the key segments of the population, regardless of the factual accuracy or, indeed, the actual benefits of a policy (Kalpokas 2019: 128). Of course, particularly in the latter case, Article 21 of UDHR, namely, that 'The will of the people shall be the basis of the authority of government', becomes noticeably less straight-forward. Instead, the key crux of governance should now be seen in the public's striving to be (or at least feel) in control, public actors' attempts to retain their ordering capacity and their progressive inability to do so without entering into hybrid arrangements with private actors, the growing role of the latter in both public and private regulation of life, datafication and disaggregation of individuals, ever-growing capacity for data-based nudge of humans embedded in algorithmically constructed environments, and the increasing autonomy of algorithms themselves.

References

Agamben, G. (1998). *Homo Sacer: Sovereign Power and Bare Life*. Stanford: Stanford University Press.

Anderson, B. (2006). *Imagines Communities: Reflections on the Origin and Spread of Nationalism* (rev. ed.). London and New York: Verso.

Bergé, J.-S., Grumbach, S., & Zeno-Zencovich, V. (2018). The 'Datasphere', Data Flows Beyond Control, and the Challenges of Law and Governance. *European Journal of Comparative Law and Governance, 5*, 144–178.

Bodo, B., et al. (2017). Tackling the Algorithmic Control Crisis: The Ethical, Legal, and Technical Challenges of Research into Algorithmic Agents. *Yale Journal of Law and Technology, 19*(1), 133–180.

Chenou, J.-M., & Radu, R. (2019). The 'Right to Be Forgotten': Negotiating Public and Private Ordering in the European Union. *Business and Society, 58*(1), 74–102.

Coglianese, C. (2007). Legitimacy and Corporate Governance. *Delaware Journal of Corporate Law, 32*, 159–167.

Cudworth, E., & Hobden, S. (2013). Complexity, Ecologism, and Posthuman Politics. *Review of International Studies, 39*, 643–664.

d'Aspremont, J. (2006). Legitimacy of Governments in the Age of Democracies. *New York University Journal of International Law and Politics, 38*, 877–917.

Edwards, L., & Veale, M. (2018). Slave to the Algorithm? Why a 'Right to an Explanations' Is Probably Not the Remedy You Are Looking For. *Duke Law and Technology Review, 18*(1), 18–84.

Foucault, M. (2012). *Discipline and Punish: The Birth of the Prison*. New York: Vintage Books.

Honig, B. (2009). *Emergency Politics: Paradox, Law, Democracy*. Princeton: Princeton University Press.

Iveson, K., & Maalsen, S. (2019). Social Control in the Networked City: Datafied Individuals, Disciplined Individuals and Powers of Assembly. *EPD: Society and Space, 37*(2), 331–349.

Jackson, J., et al. (2012). Why Do People Comply with the Law? Legitimacy and the Influence of Legal Institutions. *British Journal of Criminology, 52,* 1051–1071.

Just, N., & Latzer, M. (2017). Governance by Algorithms: Reality Construction by Algorithmic Selection in the Internet. *Media, Culture and Society, 39*(2), 238–258.

Kalpokas, I. (2018). *Creativity and Limitation in Political Communities: Spinoza. Schmitt and Ordering*. London and New York: Routledge.

Kalpokas, I. (2019). *A Political Theory of Post-Truth*. London and New York: Palgrave Macmillan.

Kreiss, D., & McGregor, S. C. (2019). The 'Arbiters of What Our Voters See': Facebook and Google's Struggle with Policy, Process, and Enforcement Around Political Advertising. *Political Communication*. https://doi.org/10.1080/10584609.2019.1619639.

Lambach, D. (2019). The Territorialization of Cyberspace. *International Studies Review*. https://doi.org/10.1093/isr/viz022.

Lessig, L. (2006). *Code: Version 2.0*. New York: Basic Books.

Levi, M., & Sacks, A. (2009). Conceptualizing Legitimacy, Measuring Legitimating Beliefs. *American Behavioral Scientist, 53*(3), 354–375.

Machin, D. J. (2012). Political Legitimacy, the Egalitarian Challenge, and Democracy. *Journal of Applied Philosophy, 29*(2), 101–117.

Mantelero, A. (2018). AI and Big Data: A Blueprint for a Human Rights, Social and Ethical Impact Assessment. *Computer Law & Security Review, 34,* 754–772.

Markham, A. N., Tiidenberg, K., & Herman, A. (2018). Ethics as Methods: Doing Ethics in the Era of Big Data Research. *Social Media+Society*. https://doi.org/10.1177/2056305118784502.

Mitchell, A. (2014). Only Human? A Worldly Approach to Security. *Security Dialogue, 45*(1), 5–21.

Muirhead, R., & Rosenblum, N. L. (2019). *A Lot of People Are Saying: The New Conspiracism and the Assault on Democracy*. Princeton and Oxford: Princeton University Press.

Riley, P. (2019). Three Pitfalls to Avoid in Machine Learning. *Nature, 572,* 27–29.

Robinson, T. D. (2017). A Normative Evaluation of Algorithmic Law. *Auckland University Law Review, 23,* 293–323.

Rothstein, B. (2009). Creating Political Legitimacy: Electoral Democracy Versus Quality of Government. *American Behavioral Scientist, 53*(3), 311–330.

Scott, L. (2019, April 21). A History of the Influencer, from Shakespeare to Instagram. *The New Yorker.* Available at https://www.newyorker.com/culture/annals-of-inquiry/a-history-of-the-influencer-from-shakespeare-to-instagram.

Suddaby, R., & Greenwood, R. (2005). Rhetorical Strategies and Legitimacy. *Administrative Science Quarterly, 50,* 35–67.

Susskind, J. (2018). *Future Politics: Living Together in a World Transformed by Tech.* Oxford and New York: Oxford University Press.

Tufekci, Z. (2019, April 22). How Recommendation Algorithms Run the World. *Wired.* Available at https://www.wired.com/story/how-recommendation-algorithms-run-the-world/.

Tyler, T. R. (2006). Psychological Perspectives on Legitimacy and Legitimation. *Annual Review of Psychology, 57,* 375–400.

Ünver, H. A. (2019). Artificial Intelligence, Authoritarianism and the Future of Political Systems. *Centre for Economics and Foreign Policy Studies.* Available at http://edam.org.tr/wp-content/uploads/2018/07/AKIN-Artificial-Intelligence_Bosch-3.pdf.

Wallner, J. (2008). Legitimacy and Public Policy: Seeing Beyond Effectiveness, Efficiency, and Performance. *The Policy Studies Journal, 36*(3), 421–443.

Weatherford, M. S. (1992). Measuring Political Legitimacy. *The American Political Science Review, 86*(1), 149–199.

Zuboff, S. (2015). Big Other: Surveillance Capitalism and the Prospects of an Information Civilization. *Journal of Information Technology, 30,* 75–89.

CHAPTER 7

Conclusion/Manifesto: Just Another Brick in the Wall? (Hint: No)

Abstract This concluding chapter deals with some of the key issues inherent in algorithmic governance: objectification, optimisation, and commodification which collectively turn humans into resources that have to be utilised with maximum efficiency. Essentially, these developments are seen as part of anthropocentrism gone wrong, ultimately tilting the balance of embeddedness—inherent to posthumanism—in favour of the digital and the platform economy. After exposing some false leads, this chapter ultimately proposes the closest thing to an emancipatory strategy available: love as a way of restoring care and repairing the balance of embeddedness.

Keywords Optimisation · Commodification · Embeddedness · Datafication · Love · Quantification

Humans have the right to be imperfect!

It should by now be clear that in both practical (algorithmic governance) and theoretical (posthumanism) terms we are embedded in our broader contexts, be them natural or digital (usually both): We 'are situated within and integrated into [our] environment' to the effect that there exists a 'dynamic interplay between human minds and the constructed socio-technical environment' (Frischmann and Selinger 2018: 91–92). Still, it has to be stressed that the environment itself is non-accidental:

© The Author(s) 2019
I. Kalpokas, *Algorithmic Governance*,
https://doi.org/10.1007/978-3-030-31922-9_7

it has, instead, been intentionally manufactured. Simultaneously, though, the manufacturing process (by technology giants, institutions, lonely great minds, etc.) is not independent but enacted in relation to other variables, such as the system of economic organisation that drives actors to act in specific ways in pursuit of specific goals (such as profits). And that system itself is simultaneously both a product and a producer of the mind, the technology, political arrangements, cultural norms, etc. which themselves are but extensions of one another and so on, ad infinitum. Hence, this book was an attempt to zoom in on a snapshot of the present stage of development within this multicausal and multidirectional process in which we are permanently embedded, by which we are permanently shaped, and which we permanently in part contribute towards shaping.

The above should already hint that suggestions such as striving for a 'practical, situated, and reasonably exercisable freedom to be off, to be free from systemic, environmentally architected human engineering' (Frischmann and Selinger 2018: 124) are neither really practical nor, strictly speaking, possible. First, the progressive datafication creep is unlikely to leave a space that is off (that would be against the internal logic of datafication) and thus there is going to be an app—and only an app—for everything. Second, and even more fundamentally, such supposed being off would mean freedom from only *some* of the causal chains. This is again where posthumanist thinking comes to the fore: we must understand our unavoidable embeddedness. In this context, the desire to break free must be seen for what it is: as yet another manifestation of the anthropocentric desire to be elevated above and independent from everything else. Hence, one thing stands out in the broader context of embeddedness and interconnection: emancipation, at least in the traditional sense, would necessitate us blasting ourselves out of this world. Instead, we need to find ways to craft better lives *within* embeddedness. That is neither a naïve nor a defeatist strategy (as e.g. Rekret [2019: 90–91] would claim)—instead, that is the sole practical solution and one that necessitates a no less radical investment.

Likewise, agency is not black-and-white either. The fake dichotomy between human and technical agency, typically implying that we can only have either one or the other fails to account for the interdependence between humans and technology 'in the collective accomplishment of affordances', the latter producing actions that are shared (Pentzhold and Bischof 2019: 7). Instead, it must be stressed that 'meaning and matter, the social and the technological, are inseparable and they do not have inherently determinable boundaries and properties; rather, they are constituted as relational effects performed in a texture of situated practices' (Gherardi 2012: 40). Problems arise when this collective accomplishment of affordances is tilted out of balance: when one element of this otherwise flat and interactive world begins to rise above all else, becoming *the* benchmark setter for any affordances. Until recently, that element was the human, basking in (usually *his*) arrogated anthropocentric glory. However, today we witness power shifting towards algorithmic rationality.

Today, perhaps the greatest human attribute to be protected is the ability to be imperfect, suboptimal, non-fully efficient. In fact, the endless quest for the optimal performance, the most efficient use possible of the resources at hand, etc. is perhaps one of the key attributes of anthropocentric folly (the quest for the ultimately and ideally superior human being), and one that has returned to haunt the human. In other words, we are now reaching a point of hubris whence humans have created powerful new tools to increase their mastery only to end up being mastered by these tools. Hence, one does not need to embark on dystopian scenarios of humans being enslaved by machines gone rogue or by some infinitely superior Artificial Intelligence—in fact, subjection is an encroaching reality, embedded in otherwise lofty-sounding goals, such as utility and efficiency maximisation, optimal performance, economic waste reduction, etc. It is, in fact, subjection as commodification that, in turn, gives rise to algorithmic governance through both architecture and nudge. This is, however, not only the commodification of humans, their lives, and their relationships—it also extends (particularly under the architectural dimension) to the built and the natural environment and, in fact, to the technology itself. As data can be, and are, gathered on everything and anything, rendering every element and aspect of this world, at least potentially, available for algorithmic analysis, optimisation, and governance, we seem to end up living in a world of universal commodity.

With universal commodification taken as a paradigm, we may well be living in what Bridle (2019) calls the 'new dark age', characterised by optimisation of humans and faith in opaque code. This optimisation is clearly evidenced in the increased capacity to quantify performance in both employment and the public sphere, creating new hierarchies therein and subjecting humans to relentless self-improvement, status work, and competition (see e.g. Mau 2019). Moreover, such quantification has a clear de-personalising aspect, which is paradoxical because *personalisation* is one of the keywords of the platform economy. However, the actual situation is this: we are at the receiving end of personalised content (although even that is a shallow personalisation, directed at us not as persons but as data subjects—as correlation supersedes causation) but the external vision of who we are becomes lost in translation to data (again, it is unimportant *who* we are but how data on us correlate with other data). In this onslaught of de-personalisation, as Mau (2019: 38) attests, 'a friendly colleague becomes a service provider, a long-trusted customer someone with a high credit score, a kind-hearted nurse a care provider, a scholar a high-impact academic, and a hobby athlete someone with a dynamic performance curve', thereby replacing personal qualities with quantifiable descriptors that denote the *usability* of the subject at hand (Mau 2019: 38). In the same vein, we become historically accumulated statuses according to which algorithmic sorting, ascription, and prediction are being performed, with the result that 'the possibilities for reinventing ourselves, changing our status or escaping it altogether are becoming fewer' (Mau 2019: 172). Once again, it is not causation (underlying personality) but data correlations that matter, emphasis being not on understanding everyday life in any deeper sense but on rendering it 'visible, readable and thereby governable, rather than seeking to understand hidden laws of causality' (Chandler 2019: 76). Therefore, this de-personalisation has, as its direct corollary, commodification: 'our data become a product that can be utilized, recombined and sold on', being made use of in ever new contexts (Mau 2019: 163).

It is also worth paying attention to the fact that optimisations and automations, so abundant today (from smart appliances to gadgets to fitness trackers to autonomous vehicles to giant systems, such as 'smart cities'), are sold on the premise of allowing us to skip some of the necessities of life in order to do more—be more efficient, more productive—to the effect that, ultimately, the physical world—from the body to geography—becomes seen as an impediment to greater productivity and to the most efficient use of, seemingly, always scarce resources, particularly immaterial ones (Frischmann and Selinger 2018: 31–33). Indeed, the sales pitch for technology-based optimisations is deceptively simple: '[t]echnology enables greater efficiency, which most people want' (Schwab 2017: 49). The question, nevertheless, is efficiency at what cost and whether that truly is what people want. Actually, upon closer inspection, it becomes evident that achievement of maximum efficiency is merely a secondary premise, dependent upon a more important ideological assumption: these technological affordances tend to 'engineer humans to behave like simple machines by treating them as resources' (Frischmann and Selinger 2018: 143). The latter needs to be stressed repeatedly: the human becomes a resource in the platform economy. Indeed, instead of public interest, the private nature of algorithmic governance deriving from the platform economy naturally privileges the treatment of humans as a resource. Definitely, it can be counterargued that posthumanism teaches us not to treat our species too seriously. Nevertheless, there is a very clearly pronounced difference: between anthropocentrism and rejection of objectification. The posthumanist logic of embeddedness rejects objectification as such, be it of humans, non-human animals, physical environment, or, indeed, the digital environment.

In fact, the legitimating discourse of the platform-based data-intensive and architecturally embedded efficiency—i.e. the gig economy— is based on some quite significant elisions. The first one is freedom. Here, Schwab's (2017: 48) assertion that '[f]or the people who are in the cloud, the main advantages reside in the freedom (to work or not) and the unrivalled mobility that they enjoy' sounds almost akin to Marie Antoinette's legendary 'Let them eat cake'. Instead, more often than not, the choice is not between working and not but between being able to subsist or not, without the social security that regular employment would bring (see, generally, Bridle 2019). In a similar way, even for those not directly participating in the gig economy, opportunities to live lives aside algorithmic architectural structuration and platform logics are growing ever fewer. Indeed, while theoretically we can, of course, opt out and not participate, '[t]he more the ecosystem turns into a global connective utilities-like infrastructure, the more citizens become dependent on that ecosystem for their private, public, and professional activities' (van Dijck et al. 2018: 149; see also Mau 2019: 4). Hence, as network effects proliferate, Schwab's (2017: 97–98) claim that 'ontological inequality will separate those who adapt from those who resist – the material winners and losers in all senses of the words' is likely to become a self-fulfilling prophecy.

Meanwhile, the second key elision is based on elimination of 'waste': Allegedly, 'by matching supply and demand in a very accessible (low cost) way, by providing consumers with diverse goods' platforms enable 'the effective use of underutilized assets – namely those belonging to people who had previously never thought of themselves as suppliers' (Schwab 2017: 20; see also McAffee and Brynjolfsson 2017: 196). In fact, McAffee and Brynjolfsson (2017: 197) straightforwardly call that 'waste reduction'. However, it is quite likely that individuals end up turning suppliers of these otherwise 'wasted' assets because there is nothing else left to supply (such as traditional labour). After all, it is far from self-evident that things like enjoying privacy and independence (as in without having to share your home or car with strangers) is actually so wasteful, and even if it is, that such wastefulness should be condemned. Indeed, as all resources are seen to have to be used as efficiently as possible, from cars to humans (hence, the spread of both self-tracking and of being tracked by one's employers), the issue is, once again, that of negative externalities, particularly objectification and commodification.

In a very similar manner, algorithmic governance, in its purest and most authoritative forms, is sold on the basis of efficiency, objectivity, and maximisation of outcomes: that whenever an algorithmic model is available, 'it tends to perform as well as, or better than, human experts making similar decisions. Too often, we continue to rely on human judgement when machines can do better' while, no less importantly, 'computers can and will come up with novel solutions that never would have occurred to us' (McAffee and Brynjolfsson 2017: 42, 116). The net result is that, as Fuchs and Chandler (2019: 12) lament, 'the human does not stand separate to or above the machine, but is contiguous with machines and inside machinic processes, as one more molecular component to be combined and recombined with others'. Of course, the posthumanist approach demonstrates that the latter embeddedness is not something entirely or unavoidably negative—instead, embeddedness in contexts, whether natural or technological, is *the* condition of the human person (and, indeed, of everything in this world). Nevertheless, the matter here is that of relative parity: that relations between mutually (or, rather, multilaterally) embedded elements are those of *interdependence* instead of *dependence*. Interestingly, humans are here receiving much of the treatment they had previously subjected their environment to. But just like human commodification of nature and the treatment of it as merely a resource the use of which has to be maximised and optimised is wrong, so is analogous placement of humans vis-à-vis machines and code.

The futility in calling for a new form of humanism to protect human exceptionality (see, notably, Frischmann and Selinger 2018: 271) should by now probably be clear. The alternative to be preferred, meanwhile, is deceptively naïve: love. This naivety is deceptive because in a world desensitised by objectification, quantification, and commodification, love is perhaps the only truly radical act. And it is love not only in the sense of 'love thy neighbour' (inasmuch as this 'neighbour' has come to be understood as the human neighbour) but love in a broader, universal sense as we need to 'extend love to include non-humans and the environment' (York 2018: 611) in order to avoid falling back into the anthropocentric trap. In other words, universal commodification must be challenged by universal love. Again, this is not some hippie stuff. It is, instead, based on the realisation that reciprocal love and care are necessary for one simple reason: everything in this world is 'fundamentally part of our own agency' (York 2018: 609). Love's simplest role here could be that of an effort amplifier: As Mickey and Carfore (2012: 129) observe, we may have many lofty initiatives for making this world fairer and more sustainable (environmentally and otherwise) but 'these efforts cannot achieve anything unless they are supplemented by love' (Mickey and Carfore 2012: 129). However, as one might expect, love exceeds this practical function.

Crucially, love is the primary relation of concern, of awe, and of unwavering support that is never intended to either decode or 'correct' difference. Part and parcel of loving others is 'the acceptance that other people are essentially different', leading a life that is 'independent and separate' (Pacovská 2017: 247). Therefore, in a loving relationship, contrary to today's trend towards datafication and de-personalising analytics, '[i]t is in principle impossible to fully understand and encompass the life of another person and to accept that is to respect the reality of another human being' (Pacovská 2017: 247). In a very similar manner but with a much broader—planetary—emphasis, '[l]ove responds to the singularity of every planetary subject, and does so with intimate contact rather than resorting to coercion, imposition, or crisis' (Mickey and Carfore 2012: 129). In fact, it is the difference and unexpectedness, openness to being surprised that undergirds love thus conceived. Perhaps even more fundamentally, this relationship is deeply interactive and formative of a human person: the experience and practice of love, from early childhood onwards, is necessary for us to be even to be capable of empathy towards others. As Tucker et al. (2005: 711) demonstrate, '[t]he exercise of empathy in both visceral and somatic domains is founded on the experience of love, both in giving and receiving'. If that is the case, then time to embrace love might be ticking out: Once optimisation and quantification takes hold from early childhood onwards (see e.g. Mascheroni 2018; Willson 2019), we as humanity may end up losing the capacity of feeling for and about others.

As is clear from the above, '[l]ove is a constructive and transformative force, not merely a passion' (Mickey and Carfore 2012: 129). And it is even more transformative in the context of datafication, platformisation, optimisation, and commodification that are synonymous with algorithmic governance through both architecture and nudge. Love as caring, love as the embrace of the unpredictable and the unknown, love as openness to the suboptimal, to the imperfect, and to the wasteful—all these facets have become radical, subversive, even revolutionary, in today's status quo. And if this love encompasses more than just the human but also living and non-living nature as well as code itself, then it is truly posthuman in the sense of going beyond narrow confines, hierarchies, and dualisms (after all, love is not love if it is hierarchical). And while love of nature is crucial, even vital, in today's world of environmental crisis and impending catastrophe, there is one more crucial bit in the preceding: we must love code. The argument of this book is not a luddite one: for all of the challenges to humanness inherent in the platform economy, technology (and platforms themselves) holds enormous potential. Hence, code is to be loved. But there is something more in the 'love code' imperative that needs to be made explicit: we need to write code based on love, code that has love instead of optimising logic at its core. Only then the crucial, the most fundamental right that must be awarded to all (humans, nature, code)—the right to be imperfect—can be materialised.

This book opened with a musical reference and thus also closes with one as the title of this concluding chapter evidently alludes to Pink Floyd's album *The Wall*. Both the album and, perhaps even more hauntingly, the film based on the music portray a world devoid of one fundamental thing: love. It is this absence of love in both a generalised and a personified sense that turns each and every one into 'just another brick in the wall'. Love only remains as a matter of 'what if?'. In most of today's interactions, particularly those mediated by online platforms, love also remains merely a matter of 'what if?'. Therefore, the most pressing challenge facing humans in their relationship with technology is how to encode love and make it *the* central architectural feature. The crux of this argument is not that such encoding would suddenly change everything ('make love not war' and so on). Humans are too complex and (yes!) imperfect. To assume otherwise would mean merely substituting one totalising ideal (optimisation) with another one. There will still be discord, there will still be emotions and affects other than love (to keep the musical theme going, why not allude to a line from an Elton John song: *Without love I'd have no anger*; indeed, in and objectified world there is only waste that has to be dispassionately eliminated). Hence, love must be embraced as open-ended and non-ascriptive—hence, non-totalising.

As a matter of final restatement, in a world relentlessly striving for quantifiable perfection, ultimate emancipation is to be found not in being more perfect than everybody else but precisely in imperfection and in loving oneself and the (human, non-human, physical, digital) other *for* their imperfection.

References

Bridle, J. (2019). *New Dark Age: Technology and the End of the Future*. London and New York: Verso.

Chandler, D. (2019). What Is at Stake in the Critique of Big Data? Reflections on Christian Fuchs' Chapter. In D. Chandler & C. Fuchs (Eds.), *Digital Objects, Digital Subjects: Interdisciplinary Perspectives on Capitalism, Labour and Politics in the Age of Big Data* (pp. 73–79). London: University of Westminster Press.

Frischmann, B., & Selinger, E. (2018). *Re-Engineering Humanity*. Cambridge and New York: Cambridge University Press.

Fuchs, C., & Chandler, D. (2019). Introduction. In D. Chandler & C. Fuchs (Eds.), *Digital Objects, Digital Subjects: Interdisciplinary Perspectives on Capitalism, Labour and Politics in the Age of Big Data* (pp. 1–20). London: University of Westminster Press.

Gherardi, S. (2012). *How to Conduct a Practice-Based Study: Problems and Methods*. Cheltenham and Northampton, MA: Edward Elgar.

Mascheroni, G. (2018). Datafied Childhoods: Contextualising Datafication in Everyday Life. *Current Sociology*. https://doi.org/10.1177/0011392118807534.

Mau, S. (2019). *The Metric Society: On the Quantification of the Social*. Cambridge and Medford: Polity Press.

McAffee, A., & Brynjolfsson, E. (2017). *Machine, Platform, Crowd: Harnessing Our Digital Future*. New York and London: W. W. Norton.

Mickey, S., & Carfore, K. (2012). Planetary Love: Ecofeminist Perspectives on Globalization. *World Futures: The Journal of Global Education, 68*(2), 122–131.

Pacovská, K. (2017). Love and the Pitfall of Moralism. *Philosophy, 93,* 231–249.

Pentzhold, C., & Bischof, A. (2019). Making Affordances Real: Socio-Materia Prefiguration, Performed Agency, and Coordinated Activities in Human-Robot Communication. *Social Media + Society.* https://doi.org/10.1177/2056305119865472.

Rekret, P. (2019). Seeing Like a Cyborg? The Innocence of Posthuman Knowledge. In D. Chandler & C. Fuchs (Eds.), *Digital Objects, Digital Subjects: Interdisciplinary Perspectives on Capitalism, Labour and Politics in the Age of Big Data* (pp. 81–94). London: University of Westminster Press.

Schwab, K. (2017). *The Fourth Industrial Revolution.* London: Portfolio.

Tucker, D. M., Luu, P., & Derryberry, D. (2005). Love Hurts: The Evolution of Empathic Concern Through the Encephalization of Nociceptive Capacity. *Development and Psychopathology, 17,* 699–713.

van Dijck, J., Poell, T., & de Waal, M. (2018). *The Platform Society: Public Values in a Connective World.* Oxford and New York: Oxford University Press.

Willson, M. (2019). Raising the Ideal Child? Algorithms, Quantification and Prediction. *Media, Culture and Society, 41*(5), 620–636.

York, M. (2018). Revolutionary Love and Alter-Globalization: Towards a New Development Ethic. *Social Change, 48*(4), 601–615.

Index

A
A/B testing, 7, 41
Action, 2, 6, 9, 12, 15, 20, 21, 29, 31,
    34, 39, 41, 43, 51–54, 57, 61,
    67–70, 72–74, 76, 78, 80, 83,
    84, 92–94, 96, 100, 108
Actionable insights, 14
Affect, 7, 22, 35, 50, 52, 55, 73, 75,
    78, 96, 97, 115
Agency, 3–5, 7, 8, 17, 31, 38, 52, 61,
    63, 67–75, 77–79, 81, 84, 90,
    108, 113
Agglomerations, 5, 8, 16, 20, 73, 76
Anthropocentrism, 79, 81, 111
Architecture, choice, 42, 57, 59, 61,
    62, 75, 77, 78
Architecture, digital, 2, 49, 50, 74, 77
Artefact, 52, 68, 70, 71, 75, 76, 82,
    92
Assemblage, 3, 16, 30, 68, 73, 74,
    79, 91
Attention, 6, 7, 19, 21, 33, 50–52, 54,
    55, 57, 77, 90, 93, 110
Automation, 15, 16, 33, 83, 99, 110
Autonomy, 6, 69, 72, 84, 104

B
Bias, 7, 32, 35

C
Capital, 3, 4, 13
Causation, 2, 74, 110
Choice, 6–8, 20, 28, 32–35, 42–44,
    49–54, 57–63, 67, 70, 72, 73,
    76–79, 82, 83, 91, 100, 102, 111
Co-constitution, 58, 71, 79, 80, 91
Code, 6–8, 27, 29–34, 36–40, 42–44,
    50, 52, 56, 58, 59, 61, 62,
    67–74, 76–79, 81, 83, 84, 90,
    93–95, 98, 100, 109, 112, 114
Cognitive bias, 60, 63
Commodification, 4, 109, 110,
    112–114
Conditioning, 5, 30, 31, 36, 41, 56,
    70
Consumer satisfaction, 12, 17, 36, 51,
    52, 61
Content, 4, 6, 7, 9, 16, 17, 28, 31,
    33, 50, 51, 61, 75, 77, 79, 91,
    99, 110

© The Editor(s) (if applicable) and The Author(s), under
exclusive license to Springer Nature Switzerland AG 2019
I. Kalpokas, *Algorithmic Governance*,
https://doi.org/10.1007/978-3-030-31922-9